Adolphe Boucard

**Handbook of Natural History**

Being an Explanation of Boucard's Series of Pictorial Diagrams....

Adolphe Boucard

**Handbook of Natural History**
*Being an Explanation of Boucard's Series of Pictorial Diagrams....*

ISBN/EAN: 9783337025632

Printed in Europe, USA, Canada, Australia, Japan

Cover: Foto ©berggeist007 / pixelio.de

More available books at **www.hansebooks.com**

# HANDBOOK

OF

# NATURAL HISTORY.

LONDON :

PRINTED BY ARLISS ANDREWS, AT THE MUSEUM STEAM PRINTING WORKS

31, MUSEUM STREET, BLOOMSBURY.

1873.

# HANDBOOK

OF

# NATURAL HISTORY:

BEING AN

𝔈𝔵𝔭𝔩𝔞𝔫𝔞𝔱𝔦𝔬𝔫 𝔬𝔣 𝔅𝔬𝔲𝔠𝔞𝔯𝔡'𝔰 𝔖𝔢𝔯𝔦𝔢𝔰

OF

## PICTORIAL DIAGRAMS & NATURAL SPECIMENS,

ILLUSTRATIVE OF

### HUMAN PHYSIOLOGY, ZOOLOGY, BOTANY,
### GEOLOGY, AND MINERALOGY.

𝔏𝔬𝔫𝔡𝔬𝔫:

A. BOUCARD,

NATURALIST, 55, GREAT RUSSELL STREET; AND

THOMAS MURBY,

32, BOUVERIE STREET, FLEET STREET.

1874.

# ADVERTISEMENT.

We earnestly request all persons engaged in tuition, who appreciate the advantages of this mode of instruction, and make use of these diagrams, to have the kindness to transmit their observations to us, both as regards the composition and arrangement of this work, and also on the method of teaching which they have adopted. Such information will enable us to correct and improve future editions; and we shall feel it a duty to acknowledge the source from whence we shall have received any advice which may aid in the improvement of this work.

A. BOUCARD

55, GREAT RUSSELL STREET, W.C.

# PREFACE.

If the study of the sciences is so much neglected in England at present, it is certainly not for the want of illustrious professors, for we may congratulate ourselves on always possessing men of the first rank in science. Nevertheless, the great majority of all classes are scarcely acquainted with even the rudiments of these branches of knowledge, because the primary instruction of children does not include any of these very useful sciences, which would be a real amusement to them, and which are of such numerous and frequent application in the arts, industry, agriculture, commerce, and, in short, the ordinary business of life.

What is to be done to remedy this state of things? We must inspire and develope a taste for the sciences from infancy, and for this purpose must select one of the simplest and most attractive practical sciences, which is also of very frequent application. Natural History will certainly answer our purpose best. In truth, the peculiar attraction which natural history possesses for children is so striking that most writers of alphabets and other elementary books, try to make them interesting to the scholars by giving pictures and descriptions of animals. Unfortunately, these pictures are often bad, and represent the rat of the same size as the lion; and thus tend to mislead the notions of children; and the descriptions are generally no better than the pictures in this respect. We were inclined to think that natural objects, or good diagrams, *of the natural size,* and coloured, would amuse the pupil, by showing him how he feels and breathes; how the grain of corn germinates; how the trunk of the oak-tree is developed; or by showing him iron and copper ores, and telling him how brass and steel are made, &c., &c., all subjects the application of which will be met with at every step in the ordinary course of life.

The first instruction for the child cannot be designed to teach many things, but ought to succeed in instilling into his mind the love of study, and to lead him to reflect. It is therefore necessary that while teaching him, we should also amuse and interest him, show him the value of knowledge, and improve his mind by awakening his intelligence.

The results of such a study are easy to foresee. The mind of the child is accustomed to compare objects with each other, and he becomes more exact in his appreciation of different things, his reason is developed, and is especially raised by the instinctive admiration which he feels for all the wonders of Creation, as he learns to know all their perfection and admirable order.

To make our work as practically useful as possible, we have made use of the observations of young children scarcely able to read, who have thus aided us, as it were, in the preparation of the book. When a word was too technical for them to understand easily, we have been obliged to change it for a common word, or to give a clear explanation; but we have also availed ourselves of the assistance of learned professors, who have kindly undertaken to revise all the diagrams in detail, and to whom the scrupulous fidelity of the representations is due.

We are then confident of having produced a work which will be understood by the ignorant, for whom it is intended; and the encouragement of professors of the highest standing gives us the assurance that it will be appreciated by the learned.

The execution of the work is based upon the following principles, which we believe to be incontrovertible:

1st. Education by the eyes is that which is least fatiguing to the intelligence and memory. In truth, when a fact is stated, and illustrated by a figure, or by a natural object, it is under-stood, and is easily fixed in the memory, and prepares it for the harder efforts of learning by heart.

2nd. Nothing is more attractive to children than the coloured representations of objects with which they are acquainted, and better still, of natural objects, when they find their name and use explained at the same time.

3rd. This education can only produce good results if all the ideas instilled into the child's mind are rigorously exact.

This work has been arranged in diagrams, composed either of natural objects, whenever this was possible, or of good coloured figures, representing with the greatest accuracy the types which we wished to illustrate, of their natural size; and we have always given the preference to those objects which come most frequently under the notice of children.

In the twenty diagrams which compose this work, we have attempted to illustrate all branches of natural history, by teaching children the most indispensable elements, or those which were most suitable to excite their curiosity, and to lead them to the desire of knowledge, from the special interest that they possess.

But that the study of these diagrams should be really instructive, some explanations besides the objects or figures were necessary; and we have supplied them as briefly and concisely as possible, so that the child may be able to read them as soon as he can spell, and we have arranged them in such a manner as to attract all possible attention, as the dia-grams cannot be looked at without reading them.

To render the work complete, we have been requested to add a handbook, in order, by means of brief, clear, and scrupu-lously exact definitions, to illustrate those points on which it was necessary, and particulars which could not be introduced into the diagrams. This book is not intended for the use of the master only, but the pupils also will be able to read it with pleasure and profit.

While always insisting on the practical side of this instruction, which is not only the indispensable introduction to the elements of agriculture and horticulture taught in the universities, but also the basis of all practical or technical instruction on general science, we have not neglected the purely scientific side, because system and classification are a great assistance to children in arranging what they are taught into an orderly series of ideas; but we have explained it in the most simple manner, so that it may be easily understood.

# METHOD OF TEACHING.

---

THE best method of teaching by means of these diagrams, is to spread them before the pupils in the course of their lessons, and then to leave them to examine them.

If they are shown the whole series at the commencement, they will look at them at first with interest, and read the names of the objects, but having much to see and to read, they will not be able to remember everything accurately, and as their curiosity is no longer stimulated by novelty, they will soon forget all.

On the contrary, by showing them those which form the subject of the lesson, they will always look upon them with great interest. Having less to read and examine at once, they will do so with more profit, and will remember them more easily; and then, when the teacher has explained to them those points which they do not quite understand, and they fully comprehend everything, the diagrams can be left at their disposal without fear. Children are fond of reading again what they already know; and the figures and names will then be firmly fixed in their memory without any fatigue, and even without knowing it.

Every time that the teacher can procure actual specimens to complete his illustrations, it will make the lesson still more profitable for the pupils by making the definitions more striking. It is often very difficult to procure the necessary types in the

animal kingdom, but specimens of the vegetable kingdom are
particularly useful, and very easy to obtain.

As regards the lessons, the best plan is to follow the manual
step by step, for it is in reality the detailed explanation of the
names and objects on the diagrams to which it refers throughout; and by thus following it, repetitions and omissions
will be avoided. But the teacher will often have to add his
own observations to what is said, and to enlarge upon the parts
which are most interesting in his neighbourhood.

Besides regular lessons, accidental circumstances will frequently give opportunities for a lesson, which should be taken
advantage of. The return of the swallows for instance will give
a good opportunity for a lesson on the migration of these birds,
and the service which they render us, like nearly all birds
which feed on insects in the spring. The children should be
forbidden to molest birds or take their nests, and be taught to
appreciate the mischief which their destruction causes at this
season of the year in particular, &c., &c.

In order to follow the lessons in their regular and proper
order, it would be necessary to begin by giving the pupils a
general idea of the value of instruction in the natural sciences;
but we must remember that the practical part of this instruction
will appear in a much more striking light to the pupils after they
have gone through the course; and it will therefore be better
not to speak of this till afterwards. In fact it is manifest that
it is extremely useful to know our organisation, and to know
by what mechanism our movements follow the directions of our
thoughts; how we breathe, see, and feel. It is not less useful
to know those animals which are real aids to agriculture, and
without which our crops would be injured, and our ruin
imminent. Alas, most of these true friends who do us nothing
but good, are generally as pitilessly destroyed as our real
enemies.

What absurd fables have been related about the poor *bat*,
which has nothing formidable about it but its reputation, and

which unceasingly pursues our enemies, the night-flying insects.

The children who destroy the nests of *tit-mice*, to put nestlings which they think they can feed on seed, into a cage, do not suspect the mischief which they are doing to the crops. These nestlings in fact can only live when supplied with plenty of insects; caterpillars, which are so numerous at the time of their birth, are their favourite food, and it has been calculated that a nest of *tit-mice* destroys about 600 caterpillars per day. If we attentively examine what each caterpillar devours, in the course of its life, we shall be able to judge how costly to agriculture are these fragile strings of small birds' eggs that children delight to make, and which their parents do not forbid them from making, because they are ignorant of the mischief which is done.

Are not *toads* often pursued, hunted out, and killed? Yet what services they render us. It is true that they were created to live in the shade, and have neither elegant forms nor brilliant colours, but they ought always to be encouraged, as they live almost entirely on slugs and injurious insects.

The study of plants is perhaps of still more general interest, and certainly of more direct utility, for they actually form the principal part of our food, and the chief source of the wealth of our country. It is therefore indispensable to learn to know them, to know how corn grows, how the trunk of the *oak* or the tuber of the *potatoe* is formed, &c., &c.; and which are the commonest edible, industrial, and poisonous plants of our country.

The earth also contains an immense store of wealth. Here are the *clay* and *kaolin* to make pottery, there *sandstone* and *flint* for paving, the manufacture of glass, &c. One country produces coal seams, the fossil remains of ancient forests buried for hundreds or perhaps thousands of years; which not only serve for warmth, but from which abundance of useful industrial products are extracted—tars, essences, beautiful red and blue dyes, &c. Another country which is marshy, possesses peat bogs, a mass of sodden vegetable débris, which when dried and prepared, forms a valuable fuel. Nearly all the products of the soil can be utilized for our requirements, and it is therefore important to know them in order to be able to use them, and to draw from them all possible advantages.

It is ignorance of these elements of science which leads to those gross errors and absurd prejudices which are really relics of barbarism, and which must be unceasingly opposed by demonstrating the simple truth, and disseminating this indispensable knowledge.

This elementary course of Natural History may be divided into about 30 lessons; and we shall rapidly point out what each of them may include; but it is obvious that they may be extended, shortened or modified, according to the time which can be given to them, and the special interest which such and such a portion may possess in connection with the district where the lesson is given, &c., &c.

*1st Lesson.*—INTRODUCTORY REMARKS — Division into three kingdoms (pp. 1—5).—Man—Races of Man (pp. 6—7).

Three diagrams, one of each kingdom, must be necessarily consulted to furnish examples.

The importance of the study of man has given him a separate and comparatively large section in the work. We have therefore gone more into particulars in this chapter than the others, and have kept it quite distinct.

*2nd Lesson.*—MAN.—Structure of the human body.—Skeleton, muscles.—Organs of digestion, of the circulation of the blood; and respiration, general observations.—Respiration and circulation (pp. 7—14, diagrams 1 and 2.)

*3rd Lesson.*—MAN.—Digestion.—Nervous system.—Organs of the Senses.—Voice (pp. 15—25, diagrams 1 and 2.)

*4th Lesson.*—ANIMAL KINGDOM.—Sub-kingdoms. VERTEBRATA —MAMMALIA, General remarks. — Quadrumana. — Insectivora (pp. 26—36, diagram 3).

5th *Lesson.* — MAMMALIA. — Carnivora. — Rodentia (pp. 36 —45, diagram 3).

6th *Lesson.*— MAMMALIA.— Pachydermata. — Ruminantia. — Marsupialia.—Cetacea (pp. 45—56, diagram 3).

7th *Lesson.*— BIRDS.— General observations.— Division into Orders (pp. 57—65, diagram 4).

8th *Lesson.*—BIRDS.—Raptores.—Scansores (pp. 65—69, diagram 4) .

9th *Lesson.*—BIRDS.—Passeres (pp. 70—77, diagram 4).

10th *Lesson.* — BIRDS. — Gallinacœ. — Grallœ.—Palmipedes (pp. 78—85, diagram 4).

11th *Lesson.*—REPTILES (pp. 86—94, diagram 5).

12th *Lesson.*—FISHES (pp. 95—103, diagram 5).

13th *Lesson.*—ARTICULATA.—Insects, General Remarks (pp. 104—109, diagram 6).

14th *Lesson.*—INSECTS.—Coleoptera (pp. 110—120, diagram 6).

15th *Lesson.*—INSECTS.—Lepidoptera (pp. 120—126, diagram 6).

16th *Lesson.*— INSECTS.— Hemiptera.—Orthoptera.—Neuroptera (pp. 126—134, diagram 6).

17th *Lesson.*— INSECTS.— Hymenoptera (pp. 134—142, diagram 6).

18th *Lesson.* — INSECTS. — Diptera. — Parasita. — Arachnida (pp. 142—150, diagram 6).

19th *Lesson.*—MYRIAPODA.—CRUSTACEA.—ANNELIDA.—INTESTINAL WORMS (pp. 151—159, diagram 7).

20th *Lesson.*—MOLLUSCA.—RADIATA (pp. 160—166, diagram 7).

21st *Lesson.*—VEGETABLE KINGDOM.—General Remarks (pp. 167—174, diagrams 8, 9, 10).

22nd *Lesson.*—DICOTYLEDONOUS PLANTS.—Umbelliferœ—Solanaceœ. — Euphorbiaceœ. — Chenopodiaceœ. — Polygonaceœ. — Papaveraceœ.—Ranunculaceæ (pp. 175—182, diagram 13).

23rd *Lesson.*—DICOTYLEDONOUS PLANTS.—Leguminosæ—Labiatœ.—Rubiaceœ.—Urticaceæ.—Lauraceœ (pp. 182—190, diagrams 12 and 15).

24th *Lesson.*—DICOTYLEDONOUS PLANTS.— Malvaceœ. — Lineaceœ. — Oleaceœ. — Rosaceæ. — Cruciferœ. —Ampelideœ (pp. 191—195, diagrams 14 and 15).

# NATURAL HISTORY

## INTRODUCTORY REMARKS.

NATURAL HISTORY relates to all those objects upon the earth which can be touched, just as Astronomy concerns all the stars which we see around us in the heavens, but which we cannot reach.

*Kingdoms.*—All natural objects are divided into three king-doms. The first includes animals, and is called the *animal kingdom.* The second includes plants, and is called the *vegetable kingdom.* Lastly, we place in the *mineral kingdom,* all objects which are neither animals nor plants; that is, those which have no life: stones, rocks, crystals, liquids, such as the water we drink, and gases, such as the air we breathe.

*Animals* have a mouth with which they eat their appropriate food. They can also run, fly, swim or walk. If we approach them, or attempt to seize and annoy them, they try to escape, or show that they feel pain. Even the oyster will forcibly close itself and resist if we try to open it, and we, therefore, say that *it feels.*

*Plants* also take nourishment, but in a different way from

B

animals. They suck up the water contained in the ground by means of their roots. They cannot move from place to place like animals, but always remain fixed where the seed fell, and took root; and lastly, if we cut off a branch from a tree, it does not seem to experience any pain; it does not feel.

*Minerals* are always easy to recognise. They have no life like plants and animals, and they do not re-produce their species like these, by young ones, eggs, or seeds.

The mineral, vegetable, and animal kingdoms supply man with everything which he requires for his food; to build houses to clothe and warm himself, or to construct tools.

The study of these three kingdoms forms what is called collectively the *Natural Sciences*, and as these concern all the beings and objects which surround us, and without which we could not live, it is clear that we ought to know them, and that the study of the natural sciences is very useful indeed.

To enable us to recognise objects among ·the innumerable number of animals, plants, and minerals which the earth contains, it is necessary to imagine an order which allows of our distinguishing every object. This order is what is called a *classification*. To attain this end, it has first been attempted to group together all those animals which have certain points of resemblance in common; all those, for instance, which suckle their young with teats have been called *mammals*; all those with feathers are called *birds*. The same with *reptiles, fishes, insects*, and *molluscs*. All those large groups of animals which have certain very important resemblances between them, have been called *classes*. We speak of the class of birds, the class of fishes, &c. But each class comprises within itself so large a number of animals that these great divisions would not be sufficient. The class of mammals, for instance, alone includes very different animals. The bat which flies like a bird, the whale which lives in the water like a fish, and the horse which runs on the ground, are all three mammals; they all three bring forth young which they suckle; and yet these three animals are not at all alike. Therefore, the animals of a class have been divided into several *orders*, including all those which have certain resemblances in common, but still somewhat distant. Lastly, in each order, those animals which have a great resemblance, though different from each other, have been put together to form what is called a *family*. Thus, the lion, tiger, and panther, are all very like a cat, and are placed in the same family. On the other hand, the family of cats feeds on flesh like the fox and wolf. The family of cats, and that which contains the fox and wolf, will both be included in the order Carnivora.

Each class, either of animals or plants, is thus divided into orders and families, in such a manner that all the beings which inhabit the earth are always arranged side by side with those that most resemble them.

Now suppose we see an animal and wish to know its history ; we shall immediately be able to find it in a book which contains the system of classification. Here, for instance, is a *pole cat*; we know immediately where we ought to look for it ; it is covered with hair; it produces young which the female nourishes; we already know that it belongs to the class of mammals; it lives on flesh, which shows us that it will be placed in the order Carnivora; and we shall soon see that it belongs to a family including the marten, the martlet, the ferret, and the weasel, all which have a long body, short legs, live in holes, and like flesh as much as the cats. We shall notice in succession the different classes, and the principal orders or families of animals.

Plants have likewise been divided into classes and families. These also are always composed of plants which have a great common resemblance, but this resemblance is not always, as in the families of animals, easily recognisable. It is generally limited to the flower and fruit.

We shall mention only the principal families, and specify the principal plants which ought to be known.

From another point of view, all animals and plants may be divided into two large groups, the *useful* and the *injurious*. The useful animals are all those which man rears for his food, for clothing, or for any other purpose. The Ox which supplies us with meat, leather, and bone, is a peculiarly useful animal; the field-mouse which devours the crops is a peculiarly in-jurious animal. Man must wage war with the latter, and he is assisted by other animals, which are themselves great enemies to injurious animals; and, therefore, all these which thus assist man, are called *indirectly useful* animals: the cat is one of these, because it eats the mice, which destroy corn, and other stores. And as the greatest enemies of man are neither lions nor wolves, nor even venomous serpents, but the insects which eat his crops, it follows that all the animals which eat insects, whether mammals, birds, reptiles, or insects themselves, are useful to man.

In order to have useful animals always at hand, man has determined to make them live with him at home. The animal is then said to be domesticated; the horse, ox, sheep, dog, cat, fowl, and duck, are domesticated animals. In other countries the elephant, and camel are also *domestic animals*.

Plants, like animals, are also divided into the useful and the injurious. They are injurious when they interfere with the growth of cultivated plants, or when they are poisonous. But at least man can always destroy them with a little instruction.

There are different kinds of useful plants. Some yield valuable medicines, like the poppy which produces opium, or the cinchona which cures fevers. These plants are called *medicinal*. Other plants are esculent, and we sometimes eat their roots, as the carrot, sometimes their leaves, as the salads; but most often their fruits. There are other plants which without being articles of food, yield what are called spices, such as pepper, cinnamon, cloves, nutmeg, parsley, chervil, garlic; there is a considerable number of these plants. Lastly there are the textile plants which yield materials which are employed to make fabrics, such as the flax, hemp, and cotton. We should never finish enumerating everything that man obtains from the vegetable world, beverages, oils, woods, dyes, and a great number of different substances, as will be seen when we proceed to the history of animals and plants, after having spoken of man.

# MAN

## RACES OF MEN.

There are several races of men which are distinguished by their colour.

Four principal races of men are recognized; the white, the yellow, the red, and the black.

The *white race* comprises those nations whose skin is more or less white, the hair silky, and the eyes blue or brown. It is the race in which civilization is most advanced. It inhabits part of Asia and Africa, and nearly all Europe. Among the blonde nations, we include the English, Swedes, Danes and Germans. Among the brown races are the Indians, Persians, Arabs, Greeks, Italians, and Spanish. The French resemble either the blonde or the brown nations around them, according to the district which they inhabit.

| White Race. | Red Race. | Yellow Race. | Black Race. |

The *yellow race* is widely spread in Asia; it includes the Cochin-Chinese, the Chinese, and Japanese, who are also civilized nations, having like ourselves been acquainted with the use of writing for a very long time. They have a yellow skin, black and straight hair, a flattened nose, and oblique eyes.

The *red race* comprises the American savages generally called (but incorrectly) *American Indians*; their skin has a reddish hue, their hair is black and straight, as in the yellow race, but they have neither oblique eyes nor a flattened nose. They are, for the most part, warlike nations who live by the chase.

Lastly, the *black race*, the most miserable of all, inhabits the whole of Central Africa, and a great part of the islands of Oceania. The skin of the negroes is entirely black, their nose flattened, their lips thick, and their hair woolly. They live in small scattered tribes, cannot write, and live by the chase. They hunt with bow and arrows, and can only build huts, while the other races, even the red race of America, have been able to raise monuments, and to make great roads.

# STRUCTURE OF THE HUMAN BODY.

SKELETON.—The body of man is supported by a solid framework called the Skeleton. The parts which compose it are called bones. There are a great number, especially in the hands and feet. The head is also composed of several small bones, but they are all soldered together, except the lower jaw, which is moveable. They form a great cavity which contains the brains. The skull has also two deep hollows in front, which contain the eyes, and which are called orbits.

In man the lower jaw is formed of a single bone, while it is composed of two parts in the sheep, ox, and a great many other animals.

When we examine a human skeleton (in which there is absolutely nothing alarming,) we see that the head is supported by a sort of column formed of massive little bones arranged one upon another. · These small bones are called *vertebræ*, and collectively the *vertebral column*. It is sometimes called the *spine*, or *back-bone*. All the vertebræ are pierced with a hole from above to below. These holes correspond to each other, and form a kind of canal which itself corresponds to the hole at the base of the skull.   This canal contains the *spinal marrow*, which is con-

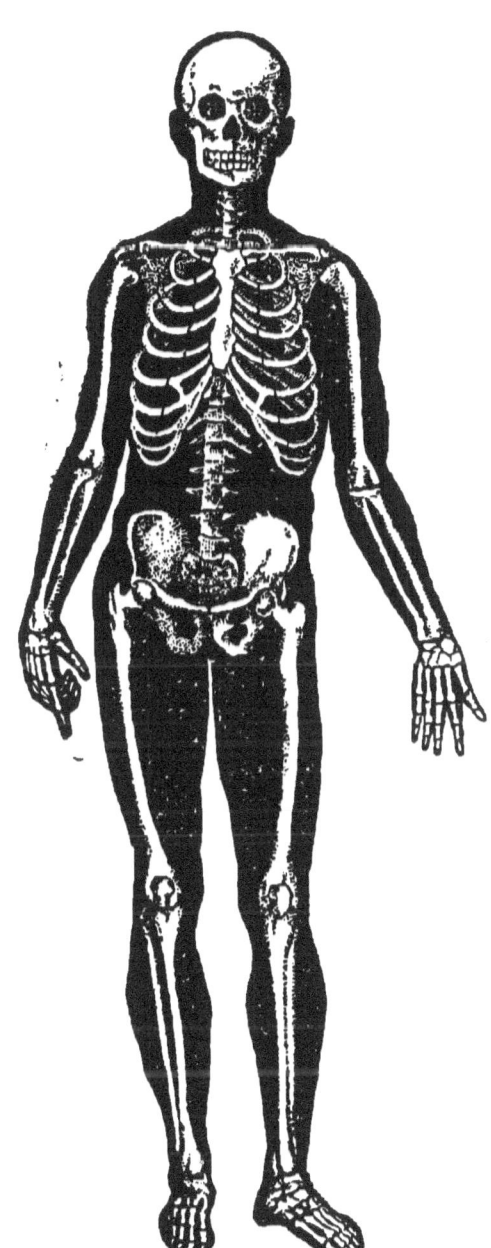

nected with the brain through the hole in the skull.   All the vertebræ are very firmly joined toge-ther.   Nevertheless it is always very wrong to lift children by the head, as is sometimes done ; for they are liable to be killed on the spot.

At the level of the chest, the vertebræ form a support for the *ribs*, which ex-tend forward, and meet against a bone placed under the skin, and called the *sternum* or breast-bone.   There are twelve ribs on each side.   Man and wo-man have therefore twenty-four.

The hips are formed by a kind of com-plete bony girdle which is called the *pelvis*. The shoulder is formed of two bones, the *clavicle*, or *collar-bone* in front, and the *shoulder-blade* behind. The clavicle can be felt under the skin above the chest on each side, and can be seen very well outlined in thin persons. The shoulder-blade is a flat triangular bone, surrounded by the flesh of the back, and it can also be seen very well in very thin people. It is not fixed, and follows the motions of raising and lowering the arms. The upper arm and thigh have only one bone; that of the thigh is called the *femur*. The fore-arm and leg have two bones placed alongside of one another; the hand and foot have a great number. The fingers and toes are divided into three parts called *phalanges*; the thumb and great toe have only two phalanges.

The bones of the limbs, for facility of movement, rotate on their extremities by means of a kind of joints called *articulations*. The shoulder, the elbow, the hip, and the knee, are the principal articulations, the phalanges are also all articulated together. The surfaces of the bone which thus slide one upon another are perfectly smooth, and in addition are always kept moistened by a sticky and oily liquid which prevents their being rubbed together too roughly.

In order to complete our study of the skeleton, we ought to speak of organs which are not so hard as the bones, and which also serve for a solid framework for the flesh; we mean the *cartilages*. The solid and elastic portions of the ear, and the sides and end of the nose are formed of cartilages. They are also formed at the extremity of all the ribs, which are osseous behind, and always cartilaginous in front.

MUSCLES.—The *muscles* form the principal part of what is called flesh. The muscles are red in man as well as in the ox and horse, but they are much paler in the sheep, the calf, and especially in the fowl. The muscles consist of fleshy masses, generally long, and continued at both ends by what are called *tendons*. The largest and best known in the body is that which ascends from the heel to the calf of the leg, and is called the *tendon of Achilles*. The calf is formed by a muscle attached above to the thigh, and which is continued below by the tendon of

Achilles, which is inserted into the bone of the heel. The kind of cords which are seen under the skin of the back of the hand when the fingers are moved, are also finer tendons, which run from the fingers to the muscles of the fore-arm. The tendons are often confounded with the nerves, and it is said for instance of a thin man making a great effort, that you can see the nerves stretch under his skin ; but this is a mistake ; it is the tendons of his muscles which are seen.

There is in the front of the arm, a well-known muscle called the *biceps*, the movements of which are very easy

to follow. To see it act well, it is only necessary to lift a tolerably heavy weight with the fore-arm ·only, by bending the elbow. The tendon which connects the biceps with the bone of the fore-arm may then be very well seen under the skin. We also perceive that the muscle contracts and thickens at the same time, in proportion as the fore-arm bends upon the elbow. This is really how the muscles act. Attached by their extremities to the bones of the skeleton, they contract at our wish, and consequently cause the bones of the skeleton to act upon each other. Each finger has also tendons which are drawn up to extend it, and drawn down to close it. Those men who have the largest muscles are generally the strongest ; but we frequently meet with very thin people who are very strong ; and they are then said to be *nervous,* owing to the same error of confounding the tendons with the nerves. We generally judge of the strength of a man by the size of the muscles of his chest, or the *pectoral muscles*. (See diagram 1.) It is these which assist in all forward motions of the arms.

# ORGANS OF RESPIRATION, CIRCULATION OF THE BLOOD, DIGESTION—DIAGRAM 1.

---

## GENERAL OBSERVATIONS.

*The neck.*—The neck contains the larynx which communicates with the throat above, and joins the *wind-pipe* below, which conducts air to the lungs.

The larynx makes a projection in man, which is felt in the neck, and is called *Adam's apple.*  When the larynx is stopped up, suffocation ensues, which is what happens in the *croup* of children.  The surface of the larynx is exceedingly sensitive ; and if it is touched by anything but air, a violent fit of coughing is the result.  This happens when we swallow anything *the wrong way,* that is when a drop of water or a morsel of food penetrates into the larynx, instead of falling into the œsophagus behind the larynx.

Below the larynx, near the wind pipe, is a gland called the *thyroid gland.*  It is not usually felt under the skin, and we only mention it because it produces *goitre,* when it swells to a large size. Goitre is very common in some countries, and seems to be connected with bad water.

*The chest.*—The chest is protected in front by the ribs, and is separated below from the abdomen by a partition called the *diaphragm.*  The chest contains the œsophagus and the wind-pipe at the back; the *heart* in front, and the *lungs* on each side. The heart is not situated wholly on the left side, as is often supposed ; the point only is a little inclined to this side ; and as it is this which is felt to beat, it was said that the heart was on the left. The two lungs fill the greater part of the chest to the right and left, but without adhering to its surface, against which they slide.

*The abdomen.*—The cavity of the belly, or of the abdomen, extends from the diaphragm to the pelvis. It is protected above only by the last ribs, and below by the hip bone. The liver is situated on the right, in the upper part of the abdomen. This secretes the *bile*, called also the *gall*, which collects in a small bladder called the gall-bladder. Further to the left is the stomach, a kind of closed bag furnished with only two openings, that of the œsophagus, by which food enters; and that of the intestine, by which it passes out. To the left of the stomach, in the upper part of the abdomen, is the *spleen*. It is there where we feel pain when we have a *stitch in the side* from having run too much. It was thought on this account that animals would be able to run faster if their spleen was removed, but this operation is no longer practised. Below the liver, the stomach, and the spleen, the *intestines* are coiled, which are at least four or five times the length of the body. They form a long tube, narrow throughout the first part of its course, which is called the *small intestines*, and larger towards the end, where it is called the *large intestines*. Behind the intestines are the *kidneys*. They secrete the urine, which drops into the *bladder* before being expelled from the body.

RESPIRATION.—It is not sufficient for man to eat to sustain life ; he must also breathe atmospheric air. This is composed of a mixture of three gases, which it is necessary to mention. The first is called oxygen, the second azote, or nitrogen, and the third carbonic acid. These three gases are mingled in very unequal proportions, and we cannot separate them at will ; but chemistry teaches us the properties of each of them.

*Oxygen* is indispensably necessary to the life of animals, as well as for the combustion of wood or coal. Where there is no oxygen, all flame is extinguished, and every animal dies. For this reason, when we are about to descend into a well, or mine, or a cistern, where no one has been for a long time, we must let down a lighted candle by a cord; if it is not extinguished, oxygen is present, and we can descend without fear; but if the candle goes out, there is no oxygen left, and a man would die there.

*Nitrogen* is a gas like oxygen, but it can neither support combustion, nor sustain life.

*Carbonic acid* is the gas which causes the froth of beer, seltzer water, cider, or of sparkling wine. Carbonic acid, like nitrogen, is neither fitted to support combustion nor to sustain life.

Air contains about one part of oxygen to three parts of nitrogen, with a very small quantity of carbonic acid. During respiration, the air which enters the lungs leaves behind a certain quantity of oxygen, and returns charged with a larger proportion of carbonic acid. Therefore, if a man is shut up in a room where the air cannot be renewed, he gradually exhausts all the oxygen, and at last dies. He dies very quickly under water, because the oxygen no longer reaches his lungs, and he can no longer breathe; and this also happens when the neck is squeezed sufficiently to compress the windpipe.

The air expelled from the lungs during respiration, contains lome aqueous vapour as well as a large quantity of carbonic acid: this forms the *moisture* of the breath, and we can thus perceive if a patient still breathes, by holding a glass to his mouth.

CIRCULATION.—The body contains a great number of vessels which proceed from the heart and return to the heart. The first are the arteries, and the second are the veins. These vessels, which grow finer and finer the further they extend from the heart, and larger and larger according as they approach it, are all filled with blood. But it is not the same colour in the veins and in the arteries; and it has no longer the same quality. It is often believed that venous blood is blue, and it

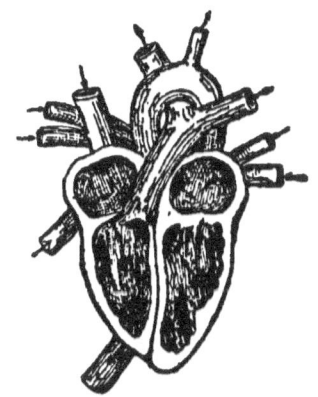

is generally represented of this colour, which is that of the veins which can be seen under the skin, either on the back of the hand or foot, or at the fold of the elbow. But when one of these blue veins is opened, as in bleeding, to draw blood from it, it is seen to be of a very dark red. It does not run with much force. If, on the contrary, a wound has unfortunately opened an artery, the blood spurts out to a distance of several yards, and it is seen to be of a vermillion red. The blood has this colour when it has been to the lungs, and has taken up the oxygen of the air derived from respiration; it loses this fine red colour in proportion as it deposits this oxygen in the tissues. Consequently, when a man is suffocated, he turns *blue*, as we say, because all his blood is of the colour of that in the veins.

The heart never ceases to beat from birth to death, in order to drive the blood into the arteries; it beats about 75 to 80 times in a minute, but sometimes much less; when it beats quicker it is a symptom of *fever*. The beatings of the heart can be counted by laying the hand on the chest, but as the pulsation is communicated to all the arteries, it is easier to feel it in those parts of the body where the arteries do not lie very deep. It is so at the wrist; and this is the place where doctors generally count the *pulse*. The beatings will be found by drawing the finger once or twice above the bend of the wrist, from the side of the palm of the hand towards that side of the fore-arm which corresponds to the thumb.

The heart is somewhat conical in shape; it contains several divisions or chambers through which all the blood successively passes. It drives the blood to the lungs, where it becomes red; and the blood then returns to the heart, which drives it through another artery into all the body. There it loses its vermillion colour, and returns to the heart by a large vein to be sent back to the lungs, and so on. This is circulation.

DIGESTION.—Man must eat and drink in order to live. All the solids and liquids which he employs for this purpose are called food But they are not all equally nourishing. It is generally necessary that a diet should be a little varied to be wholesome; but it is a mistake to suppose that one cannot live without such and such a food. In towns, one is too apt to believe that meat is indispensable to health. It may very advantageously be replaced by milk or cheese. Nor is bread indispensable, and some people eat hardly any. Habit has much to do with all this. Certainly, we may be a little inconvenienced, or even made ill, if we suddenly discontinue a diet to which we have long been accustomed, but we generally soon become accustomed, especially in youth, to a very different diet to what we had formerly been used to.

The most nourishing foods, and those which are consequently styled nutritious, are meats and vegetables. But not to grow tired of these, we generally add small quantities of other substances to them, which are not so nutritious, but which, nevertheless, greatly assist digestion, such as salt and pepper. These foods, of a special kind, have been called *condiments*.

The best and most wholesome of all beverages is undoubtedly spring water. Nevertheless, custom has almost everywhere abandoned it for the use of fermented liquors, either made of grapes, apples, or barley. Beer, cider, and especially wine, are excellent drinks, so long as they are not used to excess. But we should always be very cautious in the use of brandy, rum, and all alcoholic liquors. They have at first the serious inconvenience of causing drunkenness, in which state a man no longer knows what he is doing; but repeated drunkenness leads to much more serious consequences in time; and men who have fallen into this habit grow old before old age, their speech is confused, their hands shake, and they often end their existence in a lunatic asylum. This is the usual consequence of the abuse of brandy, and especially of absinthe, a liquor much used in France, but happily almost unknown in England.

There are other slightly stimulating beverages, very different from the preceding, such as coffee, and tea. Coffee taken in moderation is an excellent food. Tea is not, as is too often supposed, a medicine, but is very wholesome when good.

*Digestion* is the process by which food is transformed in the body into flesh and blood. Food placed in the mouth passes through the throat into the stomach and intestines, where it is digested. The mouth is always kept moist by the *saliva* which is secreted by large glands placed in the thick part of the cheeks, near the ears. It flows faster than usual when we eat, and it is enough to think of a good dinner to make the *mouth water* immediately. The teeth serve to cut, tear, and mash the food, which they form into a kind of pulp mixed with saliva. The tongue and cheeks press this pulp constantly between the teeth, till it is almost liquid. It is then only that it can be swallowed. This function of the teeth is called *mastication*. When the teeth have come out, the gums often become very hard, and we see old people who eat without teeth, nearly as well as if they had them.

*The Teeth.*—There are 28 teeth in the child, and 32 in the adult. They grow after birth, then come out, and are replaced by others which are only lost in old age. There are three kinds of teeth: the *incisors*, the *canines*, and the *molars*. The *incisors*, which serve to cut the food are the front teeth; there are four in the upper jaw,

Upper Jaw.

Lower Jaw.

and four in the lower jaw; eight in all.    On each side of
the incisors, above and below, is another tooth, stronger and
more pointed, which has been compared to those of dogs, and
is only used when we wish to tear something; these are the
*canines*, of which there are four.    As to the *molars*, they serve
to grind like mills; there are three on each side in each jaw
in children, and five in adults.

The first teeth which make their appearance after birth are
the lower incisors.    They show themselves first, and then all
the other teeth gradually appear, to the number of 24.
Towards the age of 6 years they come out, and 28 large ones
grow up in their places.    The four last which complete the
number of 32, only appear much later, at an age when one
ought to be wiser; they are called the *wisdom teeth*.    These are
the last in each row.

The teeth are formed of a very hard kind of bone which is
called ivory.    They are divided into two portions: the *root*
which is buried in the gum; and the *crown*, which is the visible
part.    This is covered with a kind of brilliant varnish, called
enamel.    In the centre of the tooth is a hole containing the
pulp; or the flesh and nerves which sometimes cause us so
much suffering.    The teeth, like the hair, should be kept very
clean, and brushed with soft brushes.    One should always
avoid breaking too hard substances with the teeth, as is sometimes done.    Without mentioning the risk of breaking a tooth,
it often happens that they crack without its being noticed, and
these teeth afterwards decay.

When anyone opens his mouth very wide, and we look down
to the back, we see behind the tongue a kind of curtain called
the *uvula*, (see diagram 2) which separates the mouth from the
throat.    On each side, below the point where the uvula com-
mences, are the *tonsils*, (see diagram 2) which very often swell
in children, impeding their respiration, and causing them much
suffering.    The part at the back of the uvula communicates
above with the openings of the nostrils; and below with the

*œsophagus* (see diagram 2) through which the food passes, and
with the *larynx*, where the air for respiration enters.  The food
passes from the throat into the œsophagus, and through that
into the stomach.  There it changes its nature entirely, and
acquires an exceedingly disagreeable taste .and odour.  We
perceive this when we *vomit;* the stomach rejects its contents,
and we can already perceive how greatly the food has been
altered.  It is still more altered when it passes into the intestines, where it is mixed with the *bile.*  Then it is absorbed by
the surface of the intestines, and is converted into blood; and
this in its turn becomes flesh, muscles, tendons, bones, cartilages, skin, hair, nails, humours; in short, all the substances which compose the various organs of the body.  What
is not thus absorbed and transformed is expelled from the body.

NERVES AND BRAIN.—The nerves are small white cords which
penetrate the whole body, and convey our wishes to every part.
If we wish to move the foot or hand, it is by means of the
nerves that our will contracts the muscles which move the hand
or foot.  We also feel by means of the nerves.  If the nerves
of the leg, for instance, have been severed by a wound, the leg
immediately becomes insensible and incapable of movement;
it is, as doctors say, paralysed.

The nerves sometimes cause great suffering, and produce
what is called neuralgia.

All the nerves in the body return to the spinal cord and to
the brain, which is a continuation of it.  The spinal cord and
the brain are composed of a very soft substance, fortunately
protected by the skull and the vertebra, for the least touch
which it sustains is always followed by the most serious consequences.  Part of this substance is grey, and the rest white.
The first forms the surface of the brain, the second is in
the centre.  The whole surface of the brain is covered with
large folds, which are called *convolutions.*

The brain is the organ by which we feel, think, remember,
or decide upon any action, such as reaching out the arm, or

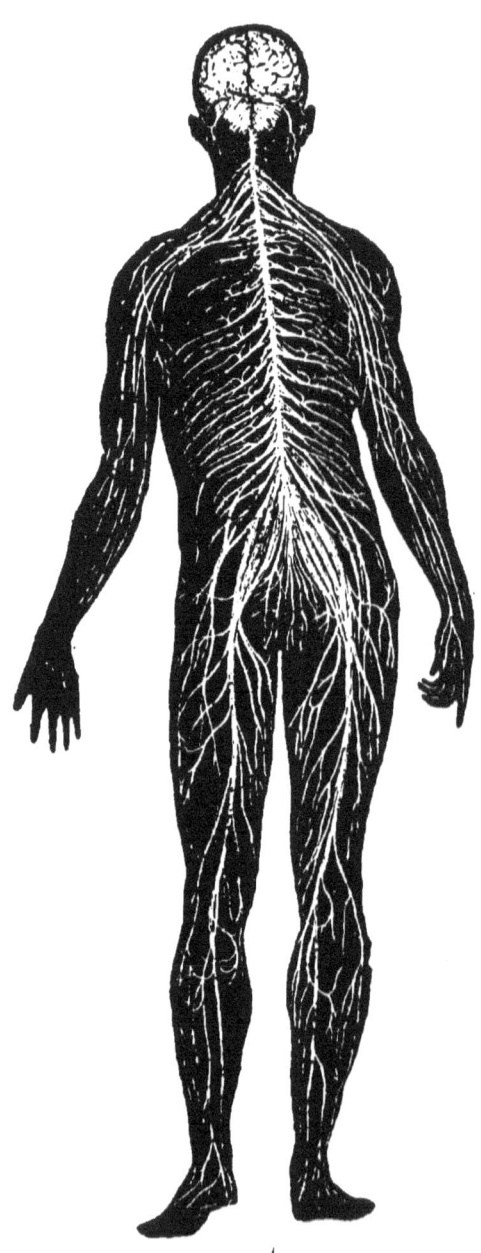

Nervous system of Man.

closing the hand. Mad-men, who are deran-ged, or who no longer know what they do, are persons whose brain is diseased. When we say that wine and brandy go to the head we are right, for they cause disease of the brain for some time, and this disease produces drunkenness. When we place our hand upon a hot or hard body, the sensation of heat or hardness is transmitted to the brain by the nerves of the skin. If we want to extend the arm, or close the hand, the nerves transmit the wish to the muscles of the hand and arm, to make the move-ment which the brain desires. The brain may be compared to a central telegraph office, connected with all parts of the body by wires, which are nothing else than the nerves.   We are informed by these wires, of everything which acts agreeably or disagreeably on the different parts of the body; and we send orders to the muscles by the same wires to make the movements which we desire.

---

# ORGANS OF THE SENSES—DIAGRAM 2.

ORGANS OF THE SENSES.—There are five senses by which we know what is passing around us : *touch, sight, hearing, smell,* and *taste.*   By touch, we ascertain if bodies are hard or soft, hot or cold, rough or smooth.   In the dark, touch also teaches us the forms of objects.   It is thus that the blind can perceive with their fingers all the objects which it is enough for us to see with our eyes, to know that they are there.   The skin is the organ of touch, as the eye is the organ of sight.

The eyes perceive the most distant objects, and inform us of their presence, even when we cannot touch them, as the clouds and stars.   They also show us the colour of objects.   The ear hears the sounds produced by sonorous bodies.   Smell informs us of the odour of surrounding bodies, when we breathe through the nostrils the air which has passed over them.   Taste is situated on the tongue and in the mouth, and the object which we wish to taste must be laid directly on the tongue itself.

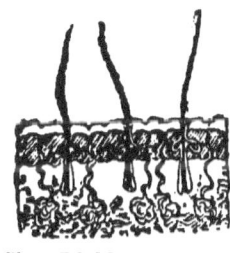

*Skin.*—The skin covers the whole body, but it is not everywhere of equal thickness.   It is especially thin on the eyelids, and especially thick on the back, under the foot, and on the palm or the hand.   It exhibits undulating lines which form elegant patterns at the end of the fingers.   These lines are separate

Skin (highly magnified).

little ridges, which we can see very well if we look with a little attention. And on these small ridges we observe rows of points like the holes which might be made by the point of a very fine needle. These are openings, and in warm weather we may observe that there is a very small drop of sweat in each of these holes. It is in reality from these that it flows.

There are other holes in the skin through which the hairs of the head and beard pass, the roots of which lie deeper. They are very seldom entirely removed when the hair is pulled out, and it nearly always grows again. The hair and beard, if left uncut, do not grow indefinitely, and after reaching a certain length, do not grow longer. The hair in children, as well as in older persons, ought always to be brushed, combed, soaped, and kept very clean. Whatever may be the current ideas on this subject, cleanliness of the heads of children is necessary to their health.

When we are slightly scalded with too hot water, or apply a blister, a portion of the skin rises, and water collects underneath. This is called a *blister*. If this skin is cut, we see that it has no feeling, and that no pain is caused. The bottom of the blister, on the contrary, is exceedingly tender, and cannot be touched without causing pain. The raised and insensible portion is called the *epidermis*, and it is only the outermost part of the skin. By working with the hands, the epidermis thickens, and then becomes *horny*. The lower part of the skin, much thicker than the epidermis, is called the *true skin*.

*The eye.*—The eye is the organ of the sense of sight. It is protected by the eyelids, and when these are closed, the eye sees no longer. It can, however, distinguish darkness and light through them, as we may perceive by standing in the sun, and bringing the hand or a dark object before the closed eyelids.

When we look at the eye of any one, we first notice a black hole in the centre, which is the *pupil*. Round the pupil is a coloured membrane in which the hole is pierced; and this is called the *iris*. The iris is blue, grey, or brown, in different persons. These tints are also sometimes slightly green, or

yellowish. If we look at the eyes of the same person in the
sun and in a dark place, we see that the pupil is not always of
the same size; it enlarges in the shade, and contracts in full
daylight. This is particularly noticeable in cats. We have
only to look at their eye in the sunlight to see the pupil reduced
to a narrow vertical line, not the tenth part of an inch in
breadth, which no longer occupies the whole height of the eye.
In the evening, and especially on a dark night, the pupil enlarges till the iris can be no longer distinguished, or is only
visible as a narrow border all round the eye.

In front of the eye is a convex transparent part extended
before the iris; this is the *cornea.* Behind the pupil, and consequently behind the iris, is an organ like a magnifying glass,
and as transparent as crystal; which is called the *crystalline lens.*
Behind this, the eye is filled with a kind of transparent jelly,
the *aqueous humour.* Lastly, the back of the eye is curtained by
an extremely delicate nervous membrane, which is called the
*retina.* It is connected with a large nerve, which runs from the
back of the eye to the brain. External objects paint themselves on the retina through the pupil, and then we see them.
If the crystalline lens grows dim, sight is lost, which happens
in *cataract.* The eye can very well be compared to the apparatus
used by photographers, and called a *camera obscura.* In front
the object-glass represents the crystalline lens and the cornea.
At the back, external objects are painted on an unpolished glass,
which is entirely analogous to the retina.

The eye, or as we say, the *ball of the eye,* moves in its orbit to
the right and left, and up and down, by means of muscles which
draw it in these four directions. When one of these muscles
is shorter than the three others, the eye is drawn to the side of
this muscle, and it is then said that the person *squints.* Everyone does not see distinctly at the same distance. Some are
obliged to hold a book very close to their eyes to read, and
others are obliged to hold it at a distance. The first are said
to be *short-sighted,* and the second *long-sighted.* Sight generally

becomes longer with age. It is generally during the first years of school that persons become short-sighted. To prevent this as much as possible, children ought not to bring their eyes too near their books and copy books; they ought to read and write holding the head straight, and at a little distance from their desk. It is the master's duty to attend to this, and the number of shortsighted children in his school will much depend on the attention which he pays to the position of the children when they work.

Too short or too long sight is corrected by means of spectacles; but the selection of these is always a matter which requires much attention, and anyone who supposes that he requires spectacles ought always to consult the doctor before going to the optician. The doctor, if he is skilful, will not only advise what spectacles should be used, but in many cases will be able to give good advice to correct the sight, and render spectacles unnecessary for the remainder of life. Tears are secreted by a gland placed in the corner of the eye, outside and above, which is called the *lachrymal gland.*

*The ear.*—The ear hears the sounds produced by vibrating bodies. It is always easy to ascertain by placing the hand on a clock when it strikes, or on the cord of a musical instrument while it is played, that bodies, when they produce a sound, experience a kind of trembling or vibration which is very perceptible to the fingers. We distinguish between the *outer* and *inner ear.* The first, visible externally, is not indispensable to hearing; it is pierced with a hole called the *auditory canal,* which penetrates into the head, and communicates with the internal ear. The bottom of the auditory canal is closed by a small membrane stretched like the parchment of a drum, and called the *tympanum.* It is therefore necessary to be always very careful not to put hard bodies into the ּauditory canal as they might break the tympanum, and cause serious accidents and dreadful sufferings.

Behind the tympanum are three very small bones of a singular form : one resembles a *hammer*, the second an *anvil*, and the third a *stirrup*. They are known by these names. Lastly we find in the internal ear a narrow canal twisted into a spiral like a snail-shell, and called, on this account, the *cochlea*.

After certain diseases, the internal ear is destroyed, and *deafness* results. If a child is born deaf, it hears nothing, and as it does not hear words, it cannot repeat and learn to say them, and is then dumb. Those who are born in this state are called *deaf-mutes*.

*The nose.*—The nose serves for respiration as well as the mouth ; and can also perceive odours. It communicates at the back with the throat, and we can therefore return smoke taken by the mouth through the nose, and can also swallow water which has been snuffed up strongly. The whole space between the nose and the throat is called the *nostrils*. They are prolonged by cavities which rise as far as the forehead, and hence it is supposed, when an irritating powder such as tobacco, pepper, or camphor has been taken, that it has penetrated to the brain ; but this is a mistake, for the brain is always separated from the nose by bones, and nothing can penetrate to it. We often speak of a cold in the head ; but it is not the brain which is affected, it is only the lining membrane of the nostrils, or the *pituitary membrane*. The brain is separated from the nose, and is not affected, and therefore cold in the head is never a serious complaint.

*The mouth* serves to breathe, eat and speak ; the *mucous membrane*, or skin which covers the inside of the mouth, also serves to taste our food. The flavour is perceived by small papilli on the tongue, each of which is connected with a nervous thread.

---

# ORGANS OF THE VOICE.—DIAGRAM 2.

The voice is formed by the air driven from the lungs, where it passes into the *larynx*. The larynx is situated above the windpipe; and it communicates with the throat by a narrow opening. It is composed of several pieces jointed together, and the interior is covered with a very fine skin, and furnished with two folds called *vocal chords*. These folds produce the sound, or voice, by being more or less tightened. The sound thus formed is articulated by the tongue with the assistance of the palate, teeth, and lips; and then constitutes speech.

# ANIMAL KINGDOM.

## CLASSIFICATION.

Sub-Kingdoms.—In classifying animals they are first divided into four large groups, called *sub-kingdoms*. These are:—

I. Sub-kingdom Vertebrata.

II. Sub-kingdom Articulata.

III. Sub-kingdom Mollusca.

IV. Sub-kingdom Radiata.

The first of these four great divisions is so named because all the animals which compose it, without exception, possess an internal skeleton; that is, a bony framework covered with flesh, like that of man, and consequently a *vertebral* column, *i.e.*, a column composed of vertebræ. This is the origin of the name of vertebrata, which is applied to this sub-kingdom. It comprises four classes : mammals, birds, reptiles, and fishes. The sub-kingdom Articulata is composed of animals whose body is formed of segments, or separate rings, arranged in a regular series. Moreover, they have no internal skeleton, and, on the contrary, the external parts are generally the hardest and toughest, as in the crayfish and the centipede. Sometimes these animals are only protected by a hard skin, like that of the earthworm, or the leech. The principal classes of this sub-kingdom are insects, crustacea, and worms.

The sub-kingdom Mollusca only contains one class, that of the molluscs. Their skin is always soft, with no appearance of rings ; the greater part are protected by a stony substance, sometimes rolled into a spiral form, as in the snail, and sometimes forming two separate parts called valves, as in the mussel.

Lastly, the sub-kingdom Radiata comprises animals which are constructed nearly like flowers, and all the parts of which radiate from a common centre. The madrepores and corals belong to this sub-kingdom.

# VERTEBRATE ANIMALS

## CLASS MAMMALIA.—DIAGRAM 3.

The first class among the Vertebrata is that of the Mammalia. Their name means "having teats." They all bring forth and suckle their young. They have generally four limbs, and are covered with hair or spines. Nevertheless there are mammals which we shall mention further on, which have no hair, and resemble fishes externally. In many of them the vertebral column extends beyond the pelvis, forming a *tail*. The number of young which mammals can produce at a birth is very variable ; the goat, the ass, the ewe, the mare and the cow have generally only one ; the hare three or four ; the dog and cat five or six ; the sow as many as fifteen.

The Mammalia are divided into several orders.

1st. The *Quadrumana*, or four-handed animals, which includes all the apes.

2nd. The *Insectivora* which are all small mammals feed on insects. In order to crush them, they have molar teeth, set with projecting points. Among the insectivora may be mentioned the bats, the moles, the shrew-mouse and the hedgehog.

Skull of hedgehog.

3rd. The *Carnivora* form an order including the large mammals which generally feed on flesh; their molar teeth are always more or less pointed in order to divide their food, and they have very strong canines to tear it.

Skull of a dog.

In the order carnivora we find the family of the bears, the badger, the family of the weasels, those of the cats and dogs, and lastly, that of the seals.

4th. The *Rodentia.*—The mammals of this order feed exclusively on vegetable matters, as the carnivora feed principally on animals. Consequently we find the most injurious mammals among the rodents. Some of them are

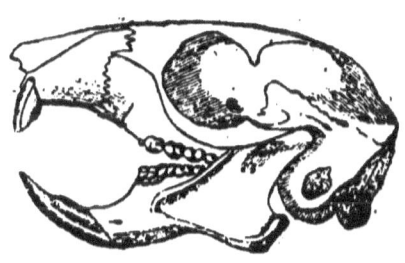

Skull of a rodent.

valued for their skins. It is sufficient to compare the teeth of a rodent, a rabbit, for instance, with those of a carnivorous animal, to see that they cannot feed in the same manner. The rodents have very strong incisors, which cut crosswise, with which they can cut wood; they have no canines, and their molars are flat to crush their food. But this is not all: the incisors are quickly worn down by cutting such hard substances; and therefore while the teeth of man and carnivora do not grow after they have acquired their full size, the incisors of rodents keep on growing all their life, as fast as they are worn away. This may be verified by cutting the teeth of a rat or a rabbit, when they will very soon regain their length.

We may mention among the rodents, the squirrels, the dormice, the moles, the marmot, the family of the rats, the field-mice, the beaver, the porcupine, and the family of the hares.

5th. Next come the *Edentata*. These are animals which inhabit tropical countries. They have no incisor teeth nor canines; and some of them have no teeth at all. They are seldom brought to Europe.

6th. The *Pachydermata* form an order which derives its name from two Greek words meaning "thick skin." Nearly all are large animals with a thick skin, and never having the feet simply cloven like the ruminants. The elephant, the rhinoceros, the horse family, the wild boar, and the hog are placed in the order of pachyderms.

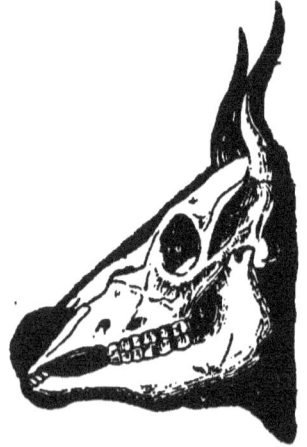

Skull of Ox.

7th. The order *Ruminantia* comprises a great many animals which have all two hoofs to each foot. Many have incisors only in the lower jaw, and none in the upper; and alone of all the mammalia, they *ruminate*. We often see a cow lying down in the fields motionless and masticating all the time, although she crops no grass. On opening her mouth, we see that she is eating afresh the food she has previously swallowed. This is rumination. Digestion is not effected in ruminants in the same way as in other mammalia; they have a very complicated stomach, or rather four stomachs between the end of the œsophagus and the beginning of the intestine. The first and largest is called the

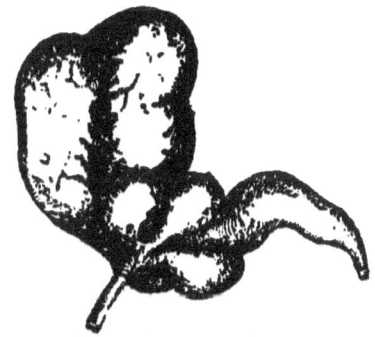

*paunch*; it is alone larger than the three others together. The second is called the *honeycomb* from its cellular appearance. The third stomach is called the *manyplus*, because its surface is lined with membranous folds. Last comes the *red*, called in calves, the *rennet*.

Stomach of a ruminant    If a piece of this is put into milk it almost immediately causes it to curdle.

This is what takes place; when the ruminating animal is in the meadow, it eats as much as ever it can, and swallows the grass almost without chewing it. All this grass goes into the paunch, where it is moistened with saliva, but does not digest. Then the animal leaves off browsing, and it is then that it really begins its meal. It returns by the œsophagus a mouthful of the grass that it has in the paunch, chews it afresh leisurely, and then swallows it; and it is only then that the food, well chewed, passes into the last stomachs, where it is digested. All animals of the order of ruminants eat thus; among others, camels, giraffes, deer, antelopes, goats, sheep, the ox, and the musk-ox.

8th. After the order of ruminants follows that of the *Marsupialia*, thus called from a Latin word which means pocket. These are mammals which are only found in the most distant countries. They are remarkable because the female has a pouch under the belly, in which she rears her young. When they are a little older, they may be seen putting their heads out of this pouch, and then drawing back and hiding there. If any danger threatens the female, she escapes carrying off her young in this manner. The best known marsupials are the opossums of America, and the kangaroos which inhabit Australia. These last mentioned animals have very short fore legs, and large hind legs,

An Opossum and young.

and instead of running, they take great leaps.

9th. The last order of mammals is that of *Cetacea*, and these are animals which at first sight have altogether the appearance of fish; such as the whale and the dolphin. They have no hair; they have fins instead of arms, and a tail behind instead of hind limbs. Nevertheless we afterwards perceive a great difference from fish. While the tail of the latter is vertical, and they beat the water on the right and left to advance, that of the cetacea is horizontal; and they move it up and down. Lastly, the cetacea have no gills; they have lungs, and breathe air like other mammals, and are obliged to return frequently to the surface to take breath. They have a nose called the blow holes, by which they blow out water, which rises in a jet from the sea.

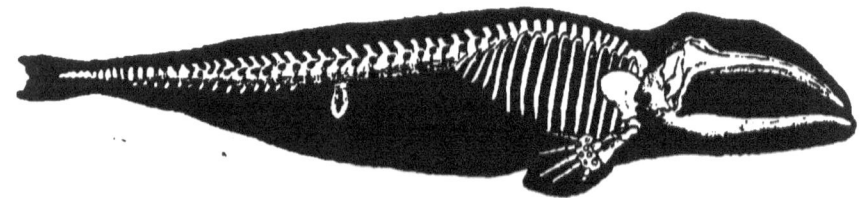

Skeleton of a Whale.

The order Cetacea includes various animals of moderate size

such as the dolphins and porpoises, which are found on our coasts ; and it also includes the cachalot and the whale.

---

# ORDER QUDARUMANA.

THE APES, which we find placed at the head of the mammals, inhabit warm countries.    The most intelligent of all is the

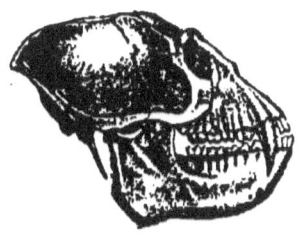

chimpanzee ; the strongest and most savage is the gorilla. Both these inhabit Africa. The gorilla is as large as a man, but its limbs are of very extraordinary strength, and can, it is said, twist the barrel of a gun.    Its teeth

Skull of ape.

are formidable as those of a lion.

---

# ORDER  INSECTIVORA.

THE BATS are the only mammals which can fly, but they achieve this with far less gracefulness than birds or insects. They are covered with hair, have

Bat flying

a mouth furnished with small sharp teeth, like all carnivorous animals, and when they are killed, the female is often found carrying her young one hanging on her shoulders, and its head downwards ; she flies everywhere with it.

Skull of bat.

Bats are nocturnal animals; they only go to seek their food in the evening, and sleep during the day. They hide in the darkest places, caves, hollow trees, and abandoned cellars; and it is probably because they have often been met with inhabiting tombs that these little animals have been considered ill-omened, and objects of fear. Nothing is more absurd. It is enough to take a bat and look at it a little while, to see that it is truly a very singular animal, but that it has nothing terrible about it except its small teeth, with which it bites those who tease it. We then see that the wing is raised by an arm, all the parts of which are visible; the arm and fore arm, at the end of which is an extremely large hand, between the fingers of which the wing is expanded. The thumb is free, and forms a kind of hook. The hind legs have also hook-like fingers, and the animal uses them to suspend itself. It clings with its sharp claws to the roof of the places which it inhabits, and if we go there in the day-time without making a noise, we shall see the bats sleeping thus with their heads downwards, hanging on all sides.

Bats have a great appetite, and when we see them flying in the twilight, they are in search of food. They only eat insects, which are injurious animals. Bats are therefore destroyers of our enemies, and far from driving them away and killing them, we ought, on the contrary, to be exceedingly glad to see them, because they are the farmer's friends, and not, as is believed erroneously, animals of ill-omen. It is therefore very wrong to kill them, and nail them to the doors of houses, where these poor animals are good for nothing, whereas they were useful when living.

We sometimes hear of a terrible bat called a *vampire*, which is said to suck the blood of men. It is true that there is in the warm countries of South America a small bat which sometimes sucks the blood of sleeping persons. But it does not take much to fill its stomach, and if the wound does not bleed after the departure of the vampire, it would not do much more harm than a leech.

Mole eating a mole cricket.

MOLES are still more insectivorous than bats, if this is possible, and have also very peculiar habits. They burrow in the ground, and make galleries in fields and meadows, and clear out the soil. This forms mole-hills. The mole lives constantly underground, and has no need to see clearly, and it is therefore nearly blind ; its eyes are not visible, as they are very small, and hidden under the fur. It uses its fore paws for digging. They are altogether disproportioned to its size, being large, great, and armed with strong claws. It burrows in the earth with this implement. There is certainly no more laborious animal. The mole sleeps very little, and works almost day and night to find its food. It is very voracious, and may be said to be always hungry. When it has not eaten for six hours, it dies of want. But it is carnivorous, and eats absolutely nothing but animals ; earthworms, wireworms, mole-crickets, and in short, all the insects that it can find. It is a serious error to suppose that it eats the roots of plants ; it dies of hunger when it has not fresh flesh to eat. The mole would thus be a very useful animal if it did not turn up the soil. In some countries, men called *mole-catchers* make a trade of destroying them, by setting traps in their galleries. In other countries they are valued, and the farmers buy them in the market to turn into their fields. Everything depends on the crops which are raised. If the field is full of mole-crickets, and if the mole-hills do not interfere with the crops, it will be an advantage to have moles ; if the earth removed by the moles causes more damage to the crop than the insects which it eats, it is better not to have moles in the field. The farmer must calculate which is best for the produce of his land.

Shrew mouse.

THE SHREW-MOUSE.—This is the small-est of all mammals.   It is smaller than the mouse ; it may be known by its much longer and more pointed muzzle, and by its teeth, which, like those of the bats and moles, are the teeth of a carnivorous ani-mal, short, sharp, made for crushing insects, whereas the mouse has teeth made for gnawing wood.

The shrew-mouse lives in the fields where it makes burrows ; it destroys as many insects as it requires to nourish its little body.   It is therefore a friendly animal, and although its aid is not of much importance on account of its size, we ought never-theless to refrain from destroying it.   It was thought that the bite of the shrew-mouse would produce a very serious disease in the feet of horses ; but this is a mistake.

*The hedgehog* is the largest of our native insectivorous animals. It destroys a great number of insects and snails of all kinds ; it does not perhaps eat so much as the mole; but at any rate it does not injure the crops.   When it is very hungry it probably eats field-mice, voles, and rats, rodent animals as destructive as insects, and which likewise appear to dread the hedgehog, as

Hedgehog.

they shun the places which it inhabits.   It passes the winter asleep in a hole.   Its skin is covered with prickles, but they would not protect it well if it did not roll itself up into a ball when attacked by an enemy. Neither head nor legs are then visible; and it remains thus until the danger is past.

---

# ORDER CARNIVORA.

BEARS.—The animals of which we are now about to speak are still carnivorous; but they no longer feed on insects. Nevertheless, if some of them are fierce and formidable animals, man has been able to turn them to profit; he hunts them for their skin, which is sometimes very valuable.

The bears are the first we come to. The white bear lives on ice in the North, and feeds on fish : the brown bear inhabits high mountains. They train it, and 'exhibit it at fairs, taking care to muzzle it well. However, the brown bear does not seem to prefer flesh; it likes fruits. It eats roots which it turns up with its claws; it is very fond of honey, and climbs trees, in spite of its apparent clumsiness, to eat bees' nests. Young bears are lively, and will play like kittens. Bearskin was long used for the fur caps of grenadiers; now this ridiculous head-dress is no longer used; their thick hide makes good blankets in cold countries. Its flesh is very good, and yields abundance of fat.

THE BADGER is closely allied to the bear, although it is much smaller. It lives in this country, and is hunted, both for its fur, and because it destroys game. When the badger is attacked by dogs, it defends itself fiercely; it lies on its back, and repels the attacks of its adversaries with teeth and claws; but by dint of numbers, they always succeed in overcoming it.

THE WEASELS.—We have now to deal with a family of true carnivorous animals which are much alike; it includes the *polecat*, the *ferret*, the *weasel*, the *ermine*, the *pine-marten*, the *beech-marten*, and the *otter*, all animals which must be mentioned.

The *pole-cat* emits a very offensive odour : it hides itself in winter in barns and granaries; and in summer it is found in hollow trees and rabbit-warrens. It is a very mischievous animal; it kills rabbits, and sometimes poultry. It darts on hares like an arrow, clinging to their neck, and never loosening its hold, in spite of their flight.

The *ferret* has long been domesticated in Africa, from whence many are brought. It is a domestic animal like the dog, but belongs to the weasel family. It sleeps almost constantly, and only rouses itself to eat. It is the most terrible enemy to rabbits ; it darts into their burrows and drives them out; but for this purpose it must be muzzled, for otherwise it would strangle them, suck their blood, and then fall asleep in their burrow.

The *weasel* is the smallest of this family, but not the least voracious. It is scarcely larger than a rat, its fur is nut-brown and the belly white. It also hides in out-houses in winter, and in summer it lives in woods, and chases birds on the bushes. It attacks young chickens, but fowls are too large for it. Sparrows are sometimes seen to assemble in troops, and drive away a weasel by flying and chirping round it. On the other hand, the weasel destroys rats and mice, so that while it is disliked in poultry-yards, it is liked in granaries, as its small size allows it to chase the rats in their holes.

The *ermine* is a little larger than the weasel, and much resembles it : it lives in northern countries. The ermine is reddish brown in summer, and is then called the *stoat*, but becomes

Ermine.

quite white in winter, when it is hunted to obtain the fur called ermine. As the animal is very small, great numbers of skins are required to make a single mantle. For this purpose the ends of the tail of the animal, which remain black at all seasons, are generally used.

The *pine-marten* and the *beech-marten* are great destroyers of eggs and poultry. They are consequently hunted. The pine-marten is known by its yellow throat, and the beech-marten by its white one. Their back is yellowish-brown, and they yield a valuable fur.

Otters live on fish, and are, so far, mischievous animals, but on the other hand, they yield a highly valuable fur. The otter is not very active on land, but when it swims it displays so much ease and agility that it is easy to see that fish cannot escape from it. Nevertheless, as it is a mammal which is obliged to breathe air, it is often obliged to return to the surface, and cannot remain long under water. Otters are often found which have been drowned in attempting to enter the weirs in search of fish, and have not been able to get out.

The *civet* is an animal found in North Africa, larger than the otter, and somewhat resembling a cat in appearance. The civet is hunted for the sake of an odoriferous substance, which is found in a kind of pouch situated near the tail.

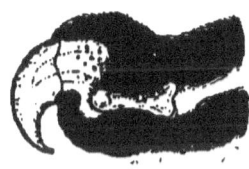

Cat's claw.

THE CATS.—The family of cats includes the *lion*, the *tiger*, the *panther*, and the *lynx*, which altogether resemble our domestic cat, except in size. They are all armed with the best teeth for tearing flesh, claws which retreat into the toes so that they cannot be blunted, and pads under the feet, which allow them to walk as noiselessly as robbers when approaching their prey.

The domestic cat is derived from the wild cat, which is found in the woods. The position which it occupies in the house is not quite the same as that of the dog. The dog never leaves it, even when he is not very well treated. The cat is more particular and more independent. It seems to have made a bargain with the master of the house, in which each is pledged to something. The cat must be fed, have a place near the fire, and full liberty to come and go, on condition of destroying the rats and mice in the house. If she is badly treated she runs away. The cat does not confine herself to the house, however well off she is there. She likes those who caress her and give her titbits, but her friendship is not proof against ill-treatment, and if she is teazed she is not slow to scratch.

The eyes of cats sometimes shine at night, and are liable to frighten children, who see only the two eyes without being able to perceive the animal. However, cat's eyes are not luminous of themselves, but only reflect the light like a mirror. If we see their eyes shine at night there must be a door or window behind us from which comes a little light, which is reflected by the eyes of the animal. During dry weather, in winter, when the cat is lying near a warm fire, we hear slight cracklings, which are also distinctly felt by the hand when we stroke her. These are slight electrical discharges, and in a dark place we can see a shower of sparks fly from the fur of the animal when stroked by the hand.

*Lions* and *tigers* hunt oxen, as the wild cat hunts rabbits. We hear of the magnanimity of the lion and the ferocity of the tiger. The truth is that these animals are more or less savage, accord ing to their personal character. We see very gentle tigers and very savage lions in menageries. Nor is the lion to be considered the king of beasts; for no animal deserves this title. The lion is neither the most intelligent nor the strongest; and the ele- phant would certainly take precedence in these respects. The lion has a mane, and the lioness has none. The tiger

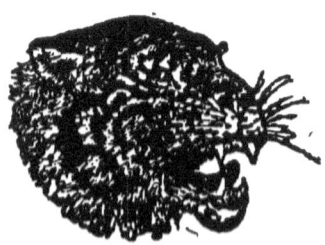

may be known by the black stripes on the reddish brown ground colour of its fur. The *panther* is smaller and is spotted. There are no lions or tigers in America, but jaguars, spotted like the panther are found there instead.

Head of Tiger.

The *lynx* is a little larger than the wild cat, and may be known by the tufts of hair at the end of its ears. It is not common except in the wilder parts of

Europe, such as Spain and Norway, and is not a native of Britain. It was formerly believed that the lynx could see better than any other animal ; and we still say of anyone who is sharp-sighted, that he is *lynx-eyed.* But its sight seems to be just the same as that of other cats.

THE DOGS.—The dog, wolf and fox form a natural family. The wolf is formidable in winter when it is hungry. It then approaches farms, and attacks the flocks, which are insufficiently guarded. In summer the wolf finds its food which generally consists of small mammals, and even carrion, in the woods. It has been extinct in Britain for the last two hundred years. The fox is celebrated for the dexterity which it displays either in creeping

Fox.

into well-secured enclosures, or in escaping from the dogs and hunters. It is also a great destroyer of poultry. When it has satisfied its hunger, it can easily carry off some dead fowls to store up in its burrow. Fox-hunting is one of the principal country sports in England.

The dog is well known to everybody, and we hardly need
• mention it. He is especially a domestic animal, and a friend of the household; he loves his master, and his friendship is proof against the worst treatment. He is intelligent, and is trained to do everything : to hunt, to guard the house at night, or to run by the carriage ; to lead the blind, and even to do errands. Dogs have been mentioned who were trained to fetch the paper for

their master every day. In some cases they are left to take care of the children, and we know how the shepherd's dog watches the flocks. In all northern countries, dogs are used to draw carriages. In Belgium and Germany four and five together are harnessed to carriages somewhat heavily laden, and others draw their master merrily along. Some nations who live in the icy North have no other beasts of burden, and fifteen or twenty are then attached to a single-sledge, and thus make long journeys across the snow.

The teeth of dogs are not so well adapted for tearing flesh as those of cats, their canines are not so long and pointed; the molars of cats are as fitted for cutting as a pair of scissors; but the last molar of dogs is flat, and formed to grind rather than to cut.

The *hyæna* which lives in Africa is considered a terrible animal, but it does not deserve this reproach. At least it is not so formidable as the wolf; it is easily tamed. It lives principally on carrion, and only attacks living animals when it is compelled to do so. As the dead are buried at a very slight depth in the country which it inhabits, it often digs up the ground to devour them, but it immediately takes to flight on the approach of a man.

The Seals.—They form a family which may be called *amphibious*, which means animals which can live either on land or in

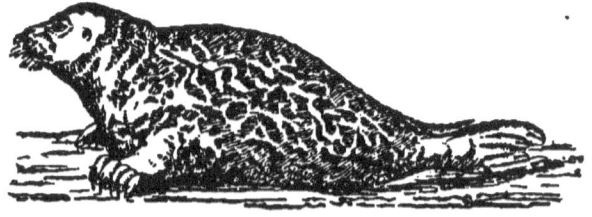

Seal.

the water. The seals are, however, easily seen to be mammals; they have fur, and four limbs armed with claws, but which they only use for swimming. Seals are found on the sea shore, where they are hunted for their blubber which yields oil, and for their fur, which is used for making tobacco pouches, caps, and a variety of other articles. When the seals are on land, they crawl along on their bellies with difficulty. They have large black eyes and a very gentle appearance. They are easily tamed, and taught to utter various sounds which have a distant resemblance to the human voice. These are exhibited at fairs under the name of *talking fish*, but they are not fish, and do not speak.

# ORDER RODENTIA.

Squirrels feed on nuts, acorns, beech-mast, &c. They build nests like those of birds among the branches of trees, large enough to accommodate all their family. These nests are made of moss and twigs, they have an opening at the top, and are protected from the rain by a kind of roof. Squirrels also lay up a store of nuts and acorns in the hollows of trees, for the winter season. They are hunted in some countries for their fur, and it is said that the hunters are skillful enough to kill them with a ball in the head, to avoid spoiling the fur.

The *dormice* are small rodents which likewise inhabit gardens and orchards, where they eat the fruit, and are consequently also very mischievous animals. They make nests like birds.

The Marmot is much larger than the squirrels; it does not climb trees, and lives in burrows. It is remarkable for sleeping all winter. There is nothing graceful about this animal, but it is very gentle. There are many of them in the mountains of Switzerland. The children catch them, and take them from village to village to show, sharing with them what is given to them. When the cold weather comes, the marmot, which has been growing fat during the summer, coils itself up at the bottom of a hole, and sleeps till spring. When it wakes, it is quite thin, and begins to eat and fatten itself again. The marmots like to live in company; they play in the meadows, but take care first to put a kind of sentinel on a rock above them, who utters a low cry when he perceives anything that might disturb the festival, and the whole band takes to flight.

⸱ THE RATS. The rat family are the greatest enemies to our dwellings. The *mouse* does less mischief than the others, on account of its small size, but it has a peculiarly disagreeable odour. There are two kinds of rats, the *black* and the *brown*. The fur of the latter is of a reddish brown. Neither are indigenous in our country, and came from Asia. Their voracity is incredible. They often eat their young ones, and if several are enclosed in a box, they eat each other till only the strongest is left; and even this has always been seriously wounded in the battles which have taken place.

Rats and mice are frequently met with which are perfectly white, and they are then called *albinos*. This name is also given to men who have white hair from youth, and red eyes. Generally they cannot bear a strong light. White mice, rats and rabbits, have also red eyes, and do not seem to see very well in broad daylight.

The *field-mice* may be known by their tails ending in a tuft of

long hair, while that of rats and mice is scaly. They are the same pests to the country that rats are in houses. However, they are not larger than a mouse, and

Field-mouse.

their fur is yellowish brown above, and dirty yellow under the belly. The short tailed field mouse lives on fruits and roots, but it prefers corn to everything else. It eats the seeds, and cuts the stalks of ripe corn; it carries to its burrow what it cannot eat on the spot, and thus stocks its small granary abundantly. Sometimes the short-tailed field-mice have been known to multiply to such an extent in a district as to become a public calamity, and to prevent any harvest being gathered in.

The *water-rat* is less injurious, but it nevertheless destroys the banks of rivers and ponds to dig its burrow.

THE BEAVER is one of the largest known rodents; it can soon cut down a tree with its teeth. It is also remarkable for its

flattened tail, covered with scales. It is celebrated for the huts which it builds. Beavers have been extinct in England for 600 years; but are still found in France on the banks of the Rhone. It only builds long burrows there; and

Beaver.

it is in the lonely rivers of North America, that it builds its villages. Several families join, and when the situation is chosen, the beavers come to shore to cut down the branches and trees which they require; they throw them into the water, and float them down to a convenient spot. Then they make dwellings of these branches mixed with earth, sometimes of a considerable size, in which they all live together. They are unfortunately becoming rarer and rarer. Beaver fur is one of the most valuable, and the hunters kill large numbers. It was long used to make beaver hats, but silk and other materials are now generally employed instead.

THE GUINEA PIG is a small rodent which is a native of South America, but which is now acclimatised with us. As it is almost defenceless, it could not live in a wild state, but it is easily reared in captivity, and it breeds very fast.

THE PORCUPINE is a rodent nearly as large as the beaver, but with the sluggish habits of the marmot. It owes its name to the fine black and white quills which grow on its back, in the place of and among the hair. Some of these animals are met with in the south of France.

THE HARES and RABBITS form one family; and everyone knows their habits. They appear at first sight to have only two incisors in the upper jaw like other rodents, but on examining them with care, two other small ones are visible behind the large ones. Rabbits breed amazingly fast when nothing interferes with their multiplication, and can spread over a whole country. The female produces from four to six litters a year; there are five or six young ones in each litter, and the young in their turn can produce young at the end of six months. It is therefore easy to calculate the rapidity with which they breed. Consequently it has been thought that it would be easy to make a fortune rapidly by breeding rabbits. But this is a great mistake, for as soon as they are much confined in a small space of country, diseases ensue which destroy great numbers.

# ORDER PACHYDERMATA.

THE ELEPHANT inhabits the East Indies and Africa. It is the largest of the Pachydermata and of all land animals. It sometimes reaches a height of 9 or 10 feet. Its strength is

Elephant.

great, and it is very intelligent. In the East Indies it is trained to fight, to hunt, and to carry very heavy burdens, which it lifts itself with its trunk, and arranges as is most convenient to it.

The elephant's trunk is simply a very long nose, which it can move at will. It breathes through two holes at the end of its trunk, which are its nostrils. There is also a small appendage at the extremity, about as large as a finger, which the elephant uses to pick up small articles. It can pick up a feather or the smallest piece of money with its trunk as easily as it can lift up and remove a cannon. Indian elephants are not generally savage, but are sometimes attacked with violent fits of rage, when nothing can resist them. They have two large teeth in the upper jaw, protruding from the mouth, and curving upwards. These are called *tusks*, and yield *ivory* which is used for so many purposes. The tusks of the Indian elephant are not thicker than a man's arm, but those of the African elephant grow to the thickness of the thigh. There is a great traffic in them. The man who guides the elephants is called *cornac* in India; he rides astride upon their neck. He pricks them, or pulls their ears with a hook to show them which way to go.

THE RHINOCEROS is another great animal which is also found in the East Indies and Africa. It does not perform the same services, and always lives in a wild state. It is chiefly remarkable for having a horn at the end of the muzzle, which is sometimes very long and pointed. Some of them have two. The substance of this horn resembles that of cow's horns, but it is solid instead of being hollow, so that a much larger quantity can be obtained from it for industrial purposes; the horn of the rhinoceros is sometimes used to make a handle for a cane, or the stick of an umbrella; but this substance is not nearly so valuable as ivory.

HORSES.—The horse family includes the *horse*, the *zebra*, and the *ass*. The horse is one of the most useful animals to man, who employs him either to draw vehicles, or to carry burdens.

Horses have only one hoof on each foot, and it is usual to add a piece of iron under the hoof to prevent it from being worn away too fast. Horses have incisor teeth in both jaws, and when they are vicious, and bite, can produce a dreadful wound.

Horse's
foot

They also defend themselves by *kicking*, either with one hind foot, or with both, and as their hoof is always shod with iron, their kicks generally produce serious wounds, and may even cause death.

There are many races of horses, which have all very different qualities. Some, like the dray and cart-horses are very good for draught; English race-horses are celebrated for their speed; Arab horses are generally small, but very hardy; they are capable of almost indefatigable exertions, and are equally proof against heat and cold; they are kept picketed out of doors. and never enter a stable. To guide the horse, a *bit* is put into

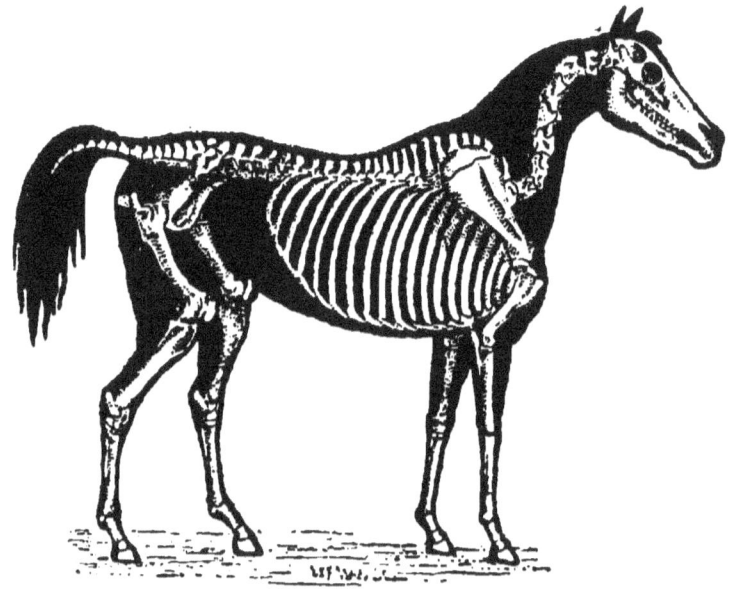

Skeleton of Horse.

his mouth, which rests on a part of the gum where there are no teeth, and which is very sensitive; so the animal stops when the bit is drawn a little tight. Old horses or those crippled by falling, are of no more use, and are killed for their leather.

The flesh is eaten in many countries, and is as wholesome as beef or mutton, to which some people prefer it.

The *ass* is very far from deserving its bad reputation, for it is a quiet, patient, and very tractable animal. When it is not ill-treated and is well fed, it does its duty zealously and cheer-·fully. It is accused of being sometimes very ·stubborn, a quality which it shares with the *mule*, which is a cross between the ass and the horse.

The *zebra* resembles the ass rather than the horse. It is covered with black and tawny stripes, which make it a beautiful animal.

The *hemionus* is also intermediate between the ass and the horse. It is smaller than the one, and handsomer than the other, and is perhaps the wild stock from which the domesticated horse is descended.

Hogs.—If there is a useful animal in the world which costs little and yields large returns, it is the hog. The wild boar, which inhabits the depths of great forests, is its nearest relation. It is armed with prominent canines, called tusks. The wild boar has four; the canines of the upper jaw rest alongside those of the lower jaw. The wild boars are fierce and savage animals; they lie all day in their retreats or *lairs*, and only go out at night to seek for fruits and roots; they dig them up with the end of their snout. When the female is about to bring forth, she abandons the male, who would eat her young ones.

The *hog* is derived from the wild boar, and much resembles it. But it is not so savage, although it has sometimes been known to devour children. It eats everything, and likes to wallow in the mud. It grunts constantly, but is nevertheless tolerably intelligent, and it has been trained to go to seek its food, and to return at a particular time. It is fattened for the table, and almost its whole body is made use of for sausages, pies, ham, bacon, lard, brawn, &c., &c.

THE HIPPOPOTAMUS is a great pachyderm which inhabits the

rivers of Africa; it has a heavy clumsy gait on land; but it swims

Hippopotamus.

in the water with great ease, and it dives and rolls about in the water with as much agility as a fish. It eats grass, leaves, and roots of trees.

## ORDER RUMINANTIA.

Camel.

THE CAMELS are ruminating animals which live in countries where there are great deserts. When they are well fed, they have one or two humps of fat on the back, which grows smaller when they are kept fasting. The dromedaries have two humps, and inhabit Asia, on the borders of Persia, in somewhat cold countries. The true camels on the contrary have only one hump, and inhabit Arabia and Africa. As these animals are able to pass several days without eating, when their paunch is full, they are extremely valuable in desert countries. But their moderation has been too much

E

praised. The camel can fast when he has not enough, but he eats gluttonously when food is abundant. It also frequently happens that he dies of hunger during the journey, and the caravan routes are strewn with his bones. The camel and dromedary supply the inhabitants of the East with milk, and wool, which is spun into clothes.

There is a much smaller ruminant than the camel, which is used for similar purposes in America. It inhabits the mountains of the Andes and the Cordilleras, and is used for the transport of merchandise. It has also an abundant fleece, which has lately been brought into use in Europe, under the name of *Alpaca*.

Giraffe.

THE GIRAFFE.—The giraffe is the largest of all ruminants, and its very long neck is terminated by a comparatively small head. The neck of the giraffe, in spite of its length, is formed of only seven vertebræ, which is the same number as in man, and in nearly all mammals, whether their neck is as short as in the elephant, or as long as in the giraffe. This animal can only browse on the leaves of trees of a considerable height, and when it wishes to take anything from the ground with its lips, it is quite a labour, and it moves its fore legs gradually apart one after the other, like some one performing a gymnastic feat, to enable its snout to touch the ground.

THE DEER are distinguished from all other ruminants by the *antlers,* which the male alone in most cases bears on his head. These antlers, in spite of their large size, are shed every year, and grow again, larger in proportion to the age of the animal. But they are not so hard as they afterwards become, when they grow. When the stag has just lost his horns, towards the end of winter, they leave two scars on the

Head of deer.

head which soon heal. The skin rises at the same time; and this is caused by the new antlers beginning to grow. Till they have reached their full size, they are covered with skin and flesh; and this skin afterwards dies and dries up, falls off in flakes, and the antlers remain, which will fall off in their turn before a year. At seven years old, the stag's antlers have ten forks, and the animal which bears them is called *dix-cors*, or *Royal Hart*.

The *fallow-deer* is smaller than the stag; and the male has much smaller horns. The hide of these animals is generally covered with white spots, which give them a very elegant appearance. They live in parks with us.

The *roe-deer* is smaller than the fallow deer, and has only very short horns. They live in families, which the members do not quit.

Head of Roebuck.

The *rein-deer* has also some resemblance to the stag. It is one of the ruminants in which the female carries horns as well as the male, though they are much smaller. As in the stag, they are shed annually. The rein-deer inhabits cold countries, where it is the only domestic animal except the dog. In winter it browses on the lichens which grow on the ground under the snow, which is enough for its support. The inhabitants of these countries use its skin and milk, and harness it to their sledges. The rein-deer has very large cloven hoofs, and does not sink in the snow.

The Antelopes form a family which includes wild ruminants, sometimes of large size, and they have true horns like oxen, which do not fall off. To this family belongs the *gazelle*, one of

E 2

the most elegant mammals in existence ; and the *chamois*, which all mountaineers delight to hunt. They sometimes risk the greatest dangers, and many. lose their lives in attempting to approach the herds of chamois. The chamois generally remain on the most inaccessible peaks, and also post sentinels who warn the herd of the approach of danger. Then the chamois escape by prodigious leaps across the precipices and rocks. It is there that they are shót, but always with balls, so that one must be very skilful, and it is an honour to kill these pretty animals, which do no harm when alive, and are worth nothing when dead. In the Pyrenees, the chamois is called *izard*.

Goat.

The *goats* are known by having the top of the muzzle straight, while it is rounded in sheep. The goat is a tame animal which yields much milk, and which is contented if it can climb on anything ; a stone, a rock, or even the branch of a tree, if it is near enough the ground. The kid yields a skin which when well prepared, is finer and more supple than any other. Gloves were formerly made of it, but kid has become very dear, and the skin of dogs and other animals are now often substituted.

There are certain goats in Asia, which yield a finer and more silky wool than the finest sheep-wool. These are the **Angora** goats. The expensive stuffs called *cashmeres* are made of their wool.

Ram.

The *sheep*.—The sheep is reared for its meat and wool. Domestication has made it weak and timid ; it cannot protect itself from the least danger, and the shepherd and his dog have always to guard the flock. Sheep are generally shorn about the month of June or July. The weight and quality of the fleece which is taken from them, vary accord-

ing to the breed; it has been known to weigh over twenty
pounds, but generally weighs ten or twelve. It is full of grease
which is removed by washing. The finest wool is the most
valuable, and is obtained from the race of sheep called *merinos.*
It is a little curled, while wools of inferior quality are harder
and stiffer. Cloth, flannel, bunting, knitting-wool, and many
other materials are made of sheeps' wool. White linen takes
the finest and richest colours in dyeing. Lastly, mutton fat is
used for soap and candles.

The *ox.*—Although the ox yields no wool, it is, like the sheep,
one of the most useful animals. It is reared for its meat, leather,
horns, and fat ; cows give their milk to make butter and cheese ;
and in many countries the oxen work, and draw vehicles like
horses elsewhere. Oxen are generally sluggish, but when
irritated they may become furious, and the sight of a red stuff
often drives them into a rage. They defend themselves with
their horns, and turn their heads to their enemies, and sometimes
toss them into the air with great violence. Oxen are not afraid
of wolves, and when they attempt to attack them, they
assemble in herds, putting the cows and calves in the
centre, and wait bravely for the wolves, or else chase them
away themselves.

It has been noticed that oxen were capable of feeling attachment, not only for those who take care of them, but also for
animals of their own species. Those which are accustomed to
be together at the plough, and know each other, do not work so
well apart, or when yoked with new comers. Sometimes the ox
is made to work with the collar like horses; and at other times a
pair is attached to the same *yoke* by the horns.

The manner in which ox hide is converted into leather, is the
same as that used for all hides which are *tanned.* The hide is
put into deep pits with bruised oak-bark, and left there for some
months. At the end of this time the hide will not rot, it has
become supple, and can be put to any known use.

Cows' milk is a no less valuable produce than beef. Butter

Cow.

and cheese are obtained from it, which can replace meat as food, and form almost the only diet in some countries. Cheese indeed supports and developes the strength of the body very well.

The quantity of milk which cows yield, varies very much according to their breed and food. Some of the breeds have long horns, others short horns, and some are hornless. The Alderney cow is particularly esteemed for the quantity and quality of its milk. Cows will yield 18 or 20 quarts of milk per day, or more. They are milked twice or thrice a day, and the milk is put into large bowls; the cream, which is the fatty part of milk, rises and swims on the surface; it is skimmed off, and beaten in a churn to make *butter*. The rest of the milk curdles; it is then put into a kind of sieve to drain. What runs off is the whey; and a solid mass is left, which is converted into the different kinds of *cheese* by various processes. But the best cheese is made of fresh milk, which still contains all the cream.

The *buffalo* is an animal closely allied to the cow, but which only inhabits the warmer parts of the Old World. It is used as a beast of burden, and is reared for its flesh, leather, and milk. The buffaloes like water, and delight to bathe in it, while the cows which accompany their herds always remain on the bank.

The *yak* is another species of ox which comes from China, and is remarkable for having a tail like a horse.

THE MUSK DEER.—We have still to mention a small ruminant animal which is found in Asia, and yields the well-known perfume called musk. It is enclosed in a pouch under the skin of the animal's belly. They kill the animal, remove the pouch with a knife, and export it at once.

# ORDER MARSUPIALIA

The marsupials include the opossums and kangaroos, which only inhabit the tropical parts of America and Oceania. These animals are all remarkable for having a pouch in front of the teats, in which the young hide when their mother escapes from danger, or they want to suck. (See page 31, for the figure of the opossum).

# ORDER CETACEA.

WHALE.—The whale is the largest and best known of the mammals forming the order Cetacea. It is hunted for its oil, and for the substance called *whalebone*, which is found in its mouth, and is used to make umbrellas and stays. Lamp-oil is chiefly composed of whale oil.

In spite of its enormous size, which is said sometimes to reach a length of 120 feet in the largest species, the whale has a very narrow throat, and can only swallow very insignificant animals. We can understand what a large quantity it must require. Therefore the whale only inhabits seas where the waters swarm with living creatures : it opens its enormous mouth, and swallows thousands of animals as large as sardines, or at most, as herrings. If its teeth were wide apart like those of most cetacea, its prey would escape. But instead of teeth, the upper jaw of the whale is furnished with hard plates close to one another, like the teeth of a comb, which allows the water to run off, while retaining the small fish and molluscs. These plates are the whalebone.

The chase of so large an animal is always a dangerous expedition. Ships called *whalers* are fitted out for the purpose, manned by hardy sailors. When they arrive at a place where they expect to find whales, a sailor is set to keep a sharp look out from the mast-head; and when they see a whale blowing, the ship is steered towards it, and boats are lowered. In front of each boat is a man with a *harpoon*, which is a dart at the end of a thick shaft of wood and iron, attached to a very strong but very slender cord, rolled on a large winch in the boat. As soon as all is ready, the harpoon is thrown, and plunged into the flesh of the whale, which takes to flight as soon as it feels the

wound; the cord runs out with great rapidity, and the rowers pull with all their strength in the same direction. It sometimes hap-

Whale fishing.

pens that the whale, thus attached to the boat, drags it to a great distance. However, it returns to the surface to breathe, and a second harpoon is thrown at it; and as soon as a good opportunity occurs, they approach near enough to thrust long lances into its body. The enormous animal is quickly exhausted, and when it is dead, it is towed towards the ship. Its fat or *blubber* makes it float. Great slices of blubber are then cut off along the whole back, which are melted, and from which the oil is extracted.

# .CLASS BIRDS

## DIAGRAM 4.

BIRDS are vertebrate animals which are always easily recognised by having only two legs and the body covered with feathers. They have also very warm blood. Nearly all birds fly, and their body seems formed to cleave the air. The beak forms a point in front, like the prow of a ship, and the body ends behind in a tail, by which the bird directs its flight, as a boat is guided by the rudder. The whole structure of the bird is arranged for flight; the feathers overlap one another like the tiles of a roof, for gliding through the air; the wings, which correspond with the arms, spread out, and fold back against the body; they are moved by very large muscles, which form the greater part of the flesh of the bird. These are those found on each side of the breast attached to a bone called the *sternum*, which is furnished with a sort of keel in the middle. Therefore the more the keel of the sternum is developed, the easier the bird flies. It is easy to see the difference in this respect between the sternum of a fowl which never rises from the ground, and that of a duck, which sometimes makes long journeys. The wings are provided with long feathers which spread out like a fan when the bird opens its wing, and fold one over the other when it closes it. These feathers are sometimes very long, and reach beyond the tail of the bird.   The longer and more pointed are the wings of a bird, the better it flies.

Skeleton of Cock.

The rump supports the feathers of the tail, which the bird moves from side to side to direct its course.   It is very easy when looking at pigeons flying, to see how they use it to guide themselves.

However, all birds do not fly equally well. There are some like fowls, which have much trouble to rise from the ground, and others, like the ostrich, which cannot fly at all. Others, instead of wings, have a kind of flat oars, with which it would be impossible for them to fly ; but which they use for swimming ; among these are the penguins and the auks.

The bones of birds are not filled with marrow, like those of the ox, and other mammals ; they are hollow and full of air, which makes them lighter, and renders flight easier.

Birds generally swallow their prey at a single gulp. The œsophagus often exhibit a fold throughout the length of the neck, well known to fanciers as the crop. When a pigeon is killed which has just been feeding, the crop is found to be filled with corn. When pigeons coo and inflate the neck, it is because the crop is filled with air. The food then passes into the *gizzard*, a stomach with a very thick shining, and almost silvery surface. The two outlets of this stomach are very near to each other ; so that it

Penguin.

requires a little attention to distinguish the œsophagus by which the food enters the gizzard, from the orifice through which it passes out.

The gizzard is nearly always found filled both with corn and small stones, which the bird swallows at the same time. The sides of the gizzard are formed of an exceedingly strong muscle ; and they contract, and bruise the corn among these stones. The product of this kind of mastication then passes into a third stomach, and into the intestine.

Birds which live on flesh instead of corn, have no gizzard, or
at least its surface is not so thick, and it does not contain any
stones.

Birds breathe like mammals by a windpipe and lungs. They
have also a larynx in the throat above the windpipe, but they
have also another in the chest, at the point where the windpipe
divides into two branches to conduct the air to each of the lungs.
It is by this second larynx that a duck can still utter a cry after
its head has been cut off. While some birds have a very
disagreeable voice, others sing, or can imitate the human voice,
like the parrot, the starling, and the jay.

Birds have the best sight of all animals; and a hawk flying at
a great height in the air can easily perceive a shrew-mouse or a
field-mouse running in the grass, and dart upon it; it is then
said that it *pounces* on its prey. Birds have generally only a
hole for an ear; but some, like the owls have a very large ear,
as large as that of a little child, hidden in the feathers on the
side of the head.

Birds' feathers are useful for a great many purposes; for
pens, for beds, and for ornament. These feathers are often very
finely coloured, and in some birds, they vary with the seasons.
Many birds have more brilliant plumage in spring than during
the remainder of the year; and the bird is then said to have
assumed its *nuptial plumage*.

All birds lay eggs. They are white in the fowl, but coloured
or spotted in other birds. We notice in the egg, 1st, the *shell*,
which is hard and re-

Germ  Airchamber

Ligament                    Yolk

Shell                       White

sistant; 2nd the *white*,
formed of albumen,
which has the pro-
perty of hardening, when heated nearly to the temperature of
boiling water; 3rd and last, the *yolk*, on which we observe a small
paler spot, which is the *germ*. The yolk also hardens when
heated. When a fresh egg has a hole carefully made in it, so
that the yolk is seen in its position, we discover that it is

enveloped in a sort of very slender skin, which forms two ligaments, floating in the albumen towards the two ends of the egg. We can also see, towards one end of the egg, a place where the albumen does not touch the shell, and which is full of air; and this is called the *air-sac*.

It is necessary for eggs to be kept at a raised temperature for some time, in order to produce chickens. In this country, the mother hatches the eggs by sitting on them, scarcely moving from them at all, until the young ones are hatched. These break the shell by pecking at it with their beaks; and still require to be brooded over by their mother for some time; and they live under her until they are grown large. The heat and care of the mother are not however indispensable to rear fowls, and they can be hatched artificially by means of an arrangement called a hatching oven, where the temperature is kept nearly equal, and sufficiently high to develop the chick. With some precautions, young chickens can thus be very easily reared.

Birds generally build nests for breeding, which are sometimes true masterpeices of architecture. Some are solidly built of  earth, others made of twigs; there are some which float on the water; and we shall mention under each species anything which is interesting about its nest. But it ought to be thoroughly comprehended thăt no nests ought ever to be destroyed, except those of birds of prey, such as falcons or hawks. All other nests ought to be respected. All young birds on coming out of the egg without exception eat insects, and nothing but insects. Even those species which destroy corn, always feed their young with

Reed Warbler.      caterpillars, grubs, and all the creatures which are most mischievous to agriculture. All who have seen young birds in the nest know what an appetite they have

They always have the neck stretched out, and the beak open, and it is all that their parents can do to provide food for this voracious family, and the quantities of insects which they then destroy, long ago caused it to be said "that there was not a single species of injurious birds in spring." For the rest, there is always a very good means of ascertaining if certain birds are useful or injurious to agriculturists; and this is to kill one or two occasionally at different seasons of the year, and to notice what food they have in their stomach; if corn, the bird is mischievous, but if it is remains of insects or grubs, the bird is useful. This is the best method of judging of the merits of such birds as rooks, which are alternately regarded as useful or mischievous. But it will not suffice to limit ourselves to examining what the bird eats once in the course of the year; it will be necessary to begin again at different seasons, because a bird which eats corn at harvest-time or seed-time for instance, destroys insects all the rest of the year; and the farmer must then calculate whether the injury done by the bird in eating his corn, is counterbalanced or not by the advantage of seeing them destroy his true enemies, *i.e.*, insects. In a general way we may say that all birds live on insects in spring, corn in summer, and berries in winter. But it must be remembered that many birds migrate during this last season.

There are, in reality, a great number of birds which are accustomed to change their country according to the season, and to make what is called a *migration* every year. Thus, when insects begin to disappear at the first cold weather, all the swallows depart for Africa, from whence they return in the spring of the following year. These long journeys are very common among birds. The cold drives them all towards the south. Those of the north come to us during the winter in search of water which is still unfrozen, and our own birds migrate to the south in search of warmth and insects. On crossing the Mediterranean Sea from France to Algeria, it is common to see flocks of small birds alight on the masts of ships, and rest there for some time before resuming their long journey, on which many doubtless perish.

---

# ORDERS OF BIRDS.

Birds, like mammals, have been divided into a number of families, in which those most alike are placed together. The following are the principal orders :—

1st. *The Birds of Prey, or Raptores.* It includes birds which are all carnivorous. They are known by their beak, which is always very strong, short, and hooked, for tearing flesh. They generally fly very well. Lastly, their toes are free in their movements, and armed with powerful claws

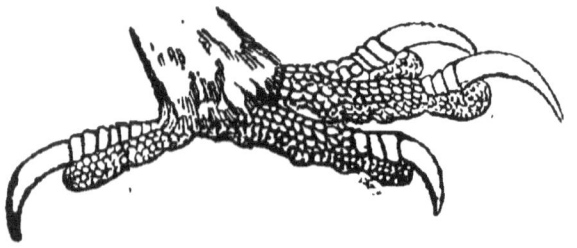

Foot of Eagle.

called *talons.* The order of the birds of prey includes the families of the falcons, vultures, and owls.

2nd. *The order of Perching Birds* includes birds generally of small size. But they may best be known by always having very strong legs fitted for grasping the branches of trees, with two or three toes directed forwards, close together, and almost united, while the other toe or toes are directed backwards, opposite to the former. Among the Perching Birds we shall mention the families of the parrots, the cuckoo, and the woodpeckers.

Foot of Parroquet.

3rd. The *Order of the Finches.* Under this name we include the greater number of the small birds which feed on either grain or insects, and which all fly well. They prefer to rest on trees, rather than on the ground; but there are some, like the lark, which run very well. The wagtail also walks gracefully, but the greater part can only advance on the ground, like the sparrow, by a series of little jumps, called *hopping.* Nevertheless some birds have been placed in this order which are very different from these; and we class among the finches the crows, the birds of paradise, the families of the warblers, the sparrows, the goatsuckers, the swifts, the swallows, and the kingfishers.

4th. *The order of Gallinaceous Birds* includes a great many birds which fly with some difficulty, always excepting the pigeons. Some are reared as domestic animals; others are valued as game. We shall mention the turkey, the peacock, the cock and hen, the grouse, the partridge, the quail, the guinea-fowl, the pheasant, and the pigeon.

5th. *The order of Waders.* In this order are arranged birds which have generally long legs, so that they seem to walk on stilts, such as the ostrich, the cranes, the heron, and the stork. Some smaller birds are put with them, such as the snipes, the woodcocks, the water-hens, the ruffs, and the lapwings, all of which have very long legs for their size. All the birds of this order are swift runners.

6th. *The order of Water birds or Web-footed birds.* This order includes all the swimming birds which have *palmated* feet; that is with the toes joined by a membrane which converts them into a kind of oar, by means of which the birds swim on the water, or even under water, for there are many of them which dive, and pursue the fish on which they feed beneath the surface. Among the swimming birds may be mentioned the gulls, the cormorants,

Foot of Duck.

the pelicans, the ducks, the swans, the geese, the penguins, and the auks.

---

# ORDER OF BIRDS OF PREY, OR RAPTORES.

FALCONS.—The family of Falcons also includes the eagles, the hawks, the sparrow-hawks, the kites, and the buzzards. All are formidable animals to rabbits, partridges, larks, and the various small birds which eat insects; and they are therefore enemies which always ought to be destroyed. They are called *birds of prey* in the strict sense ; they have a hooked beak, and very strong and pointed claws, called talons, with which they seize their prey ; and they kill them, and tear them to pieces with their beak. In some countries falcons are still used for the chase ; and falconry used formerly to be a very favourite amusement in England. The falcons are trained without too much difficulty, and then carried to the chase on the wrist. Their head is covered with a hood which prevents their seeing : and when the game is in sight, the falconer takes off the falcon's hood, and shows him the prey. Other birds, such as the heron, can be chased by the falcon ; or hares, and even larger animals ; the

Falcon hooded.

E

pursuing falcon darts upon them, and splits their skull with a blow of its beak. The falcons generally used for the chase are the jerfalcon, and the peregrine falcon. The last, which is commoner, is nearly as large as a buzzard.

The *eagles* only inhabit mountainous countries; and they generally make their nests among the rocks. These are constructed of branches roughly heaped up on the ground; and this is called the eagle's eyry. Some eagles are strong enough to carry off lambs, and could even carry off little children. This, although it has sometimes happened, is fortunately very unusual.

Head of Eagle.

VULTURES.—The vultures' inhabit warm countries, and have only been noticed in Britain on one or two occasions. They can be attracted from a great distance by the smell of carrion. They do not usually feed on fresh-killed prey, and eat only dead animals. Some are very large. They have a hooked beak like the falcons, but not so strong; their claws also are less curved, and they settle more frequently on the ground. They are remarkable for having the neck bare of feathers. To the family of the vultures belongs the *condor*, which has the highest flight of all birds, and is seen to soar above the highest mountains of America.

Head of Vulture.

OWLS.—The owl family includes several species which have a great general resemblance. These are also carnivorous birds, as is shown by their curved beaks, and talons like those of falcons;

but for all that, they are friends rather than enemies to man. They live near buildings, and actively pursue field-mice, and other small quadrupeds. A brown owl can readily take the place of a cat in a house ; and no more mice will be seen there.

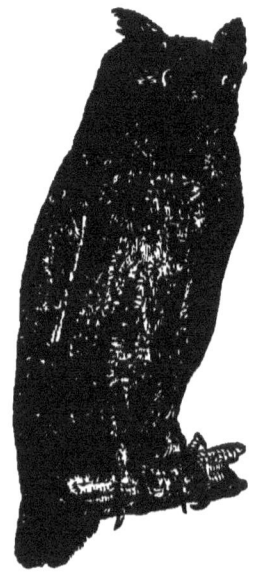

They also eat many insects which only fly by night. All these birds have an easily recognisable appearance; their two large eyes are placed in front, instead of on each side of the head, as in other birds. They have often tufts resembling ears on the head. Their ears are very large, as we have said, but it is necessary to part the plumage in order to see them. Owls, like many other animals, can see by night, and probably better than during the day, when they shun the light. They then hide in holes, and it is doubtless their habit of living in deserted places, such as cemeteries, which has led them to be regarded as birds of

Brown Owl.

ill-omen. In truth, there is no animal which deserves to be so regarded. And it is also a gross error to suppose that the owls come to hoot over a house where a dying person is lying. If we hear of it sometimes, it is because the trouble that has come upon the house keeps everyone awake, and they hear the bird's hoot, as they do every night, only on other nights everyone is asleep, and so nobody hears it. In former times, the owl, instead of being regarded as a bird of ill omen, was considered one of the wisest of animals ; and it deserves this reputation as little as the other.

# ORDER OF PERCHING BIRDS.

PARROTS.—All these birds come from distant countries; but their beautiful colour, their intelligence, and the ease with  which they learn to speak have made them valued among us. They fly badly, and feed on corn, which they break into small pieces with their beak before swallowing it; and on fruit, which they take in their claw. Their tongue is fleshy, instead of being hard and horny, as in other birds.

**Head of Parrot.**

The *cuckoo.*—The cuckoo migrates in winter, and only passes the summer with us. It is found in woods, its back is ashy, and its belly white, with fine black and grey streaks: its plumage is something like that of the sparrow-hawk, but it is easily distinguished from it by having its toes close together, two before and two behind. The cuckoo eats a considerable number of caterpillars, but it owes its celebrity chiefly to its habit of making other birds hatch its eggs and rear its young.

The female lays two eggs in the space of two or three days. She lays them anywhere upon the ground. She then immediately takes the egg in her beak, and puts it in the nest of some other bird, generally choosing one smaller than herself. But she does not abandon it, and if she sees that the bird neglects her egg, she takes it away, and puts it into another nest. When the young cuckoo is hatched among the family where it has thus been placed, it begins to try and get rid of the other young ones. By means of its rump and wings, it creeps under them, lifts them on its back to the edge of the nest, and throws them down, so that it alone remains to take the food which the owners of the nest bring to it without seeming to notice that their young ones are replaced by this stranger. This has given rise to the expression, as ungrateful as a Cuckoo.

The *woodpeckers* are insect eaters, and excellent tree climbers. They are known by their straight, strong beak, and by the feathers of the tail, which are always worn at the end, because they rest on them. Their hooked claws cling to the bark, and allow them to run along the trunks of trees, and even under the

large branches. Their plumage is sometimes beautifully coloured. They are naturally wild and they pass their lives in constant activity. Their tongue is of extraordinary length, and can be thrust out of the beak to a great distance ; they bury it under the bark, and in the holes of the wood, to seize the insects which hide there. The woodpecker is also accustomed to strike the trunks of trees sharply with its beak, in order to drive out the insects. After each blow it runs round the trunk to

Woodpecker perching.

see if it has succeeded in driving out any grubs or caterpillars from under the bark. It has been stated that it did so after each stroke of its beak, to see if it had pierced the tree from side to side. This is a fable, like so many which have been invented by those who did not fully comprehend the actions of animals.

# ORDER OF FINCHES.

The *kingfishers* are finches, they have a straight beak like the

woodpeckers, and three toes in front, two of which are partly united. Their food consists of aquatic animals. They are also brilliantly coloured. Their patience is

Foot of Kingfisher. extraordinary, and they are often seen sitting motionless on branches or stones at the edge of the water, watching for what may pass, and darting like an arrow on the prey which they perceive. Sometimes, too, they fish flying, and then, pouncing into the water, they rise again immediately with the animal which they pursued in their beak. They make their nests in the holes of the banks, only consolidating the sides. They lay from four to eight eggs, which are generally white. The male and female sit on the eggs alternately, and share the labour of feeding the young by bringing them the results of their fishing.

The *goat-suckers* are remarkable for the enormous size of their beak when open, though the horny part of the beak is small. Their plumage is dull-coloured. Many absurd stories have been told of these birds. It was believed, for instance, that they come to suck goats, whereas they only come to search in the hair and wool of sheep and goats for the insects that are found there, and of which they relieve them. As they live on no other

Head of Goat-sucker.

food than mosquitoes, gnats, and all kinds of twilight-flying insects, the goatsuckers are really very useful birds, which ought on no account to be destroyed. They do not pass the winter with us;

they migrate in autumn, when their food begins to grow scarce. Their nest is nothing but a convenient hole at the foot of a tree, or in a rock, or even in the midst of paths in woods.; and they lay two eggs marbled with bluish spots on a grey ground.

The *swifts* seem only to enjoy themselves in flight; they have very long wings, and such short legs, that they have the greatest difficulty in walking. On the other hand, as their claws are very sharp and slender, they cling easily to walls.

The swift reaches this country in the spring, a little before the swallows, and leaves a little sooner than the latter. It builds its nest under the eaves or projecting roofs of houses, as well as in cracks in walls and rocks, but when it finds the nest which it built the year before, it does not take the trouble to build a new one. It lines its nest with feathers which it has found floating in the air, or which it picks up off the ground, and which it sometimes steals from the nests of other birds, and especially from the sparrows. Sometimes, too, the swift, instead of building a nest, is contented to repair that of another bird which it adapts for its use, and where the female lays three or four white eggs.

The *swallows.*—The beak of the swallow appears to be very small, but is cleft to the eyes. They live on insects which they catch flying, and assemble together in large flocks.

About the beginning of April the swallows are seen to return to the nests which they constructed in former years. On the approach of winter, they assemble in multitudes on the roofs or trees, and after a great fuss accompanied with cries like the tumult of a debate, they start off on their journey of some hundreds or even thousands of miles.

No bird appears to fly with so much ease as the swallows; they eat, drink, and sometimes even feed their young on the wing. They are chiefly insectivorous, and consequently render great service to agriculture. Their nests are generally built against walls or buildings. They are cemented with earth in the angles of walls, or eaves, with a small opening for the family to go in and out. All the swallows which live in the same place appear to love their society, and render mutual assistance when necessary, either to repair a nest which has been partly destroyed, or to drive away a sparrow, which believing himself stronger, has come to steal something from it; all the swallows begin to harrass him, and the robber is soon obliged to fly. The swallows take great care of their young, and feed them so well that they sometimes weigh as much as their parents; and then they teach them to fly. To entice the young to fly and to leave the nest, the parents sometimes hold up before them in their beak some insect of which they are very fond.

There are two principal kinds of swallows in this country. The *house martin* is pure white on the lower part of its body, and the upper part is of a shining black with blue reflections; it

is less familiar than the swallow which has a dark band across the chest, and a much longer tail, and it arrives a little later. The latter makes its nest even in stables, under sheds, and is sometimes found living in smithies above the anvil, without the noise of the hammers, and the red sparks, seeming to frighten it.

Swallow.

The *humming-birds* are at the same time the smallest of all

birds, and those with the most brilliant plumage. They display metallic reflections of all hues, yellows, blues, greens, violets, and reds. But the humming birds do not live

Humming bird, **natural size.**

in this country, and are only found in the hot countries of America. The smallest species lay eggs scarcely larger than a pea. The humming birds are constantly on the wing; they are courageous animals, and are not afraid to defend themselves against much stronger, but less agile enemies.

The *birds of paradise* are also very beautiful birds, which are found in New Guinea, and the neighbouring islands. They are nearly as large as magpies. They are hunted for their feathers, which are made into ornaments for the toilet. The savages who sell birds of paradise to the merchants, were formerly accustomed to remove their legs, to make believe that they never rested on the ground or in trees. But the birds of paradise have legs like all other birds, which are in fact, rather ugly.

Birds of Paradise.

*Crows.*—The family of crows includes several kinds of birds which are found in England; the large hooded crow, black, with the back and belly grey; the rook, black with blue reflections, and the base of the bill bare of feathers; the jackdaw, smaller than the others, with the upper part of the head ashy grey; and lastly the jays and the magpies. All these birds have a strong beak with a cutting edge, and they are nearly all of dull colours, like birds in mourning. It is probably for this reason only that they have also been looked upon as birds of ill-omen. This belief is as absurd as all others of the same class. All these birds are generally intelligent; they are easily reared; they like the house, and learn also to repeat some words.

The *raven* is a large bird, wholly black, somewhat rare, and only found in thinly peopled districts. They live solitary in pairs, and nest on trees, or in the holes of rocks. The outside of the nest is made of the roots and branches of trees, and it is lined inside with moss or grass. The female lays five or six eggs in March; they are pale bluish green, with blotches. The male helps to sit, and to rear the young. The ravens are courageous, and are not afraid of either cats or dogs; they are attached to their master, and have been known, after having left the house to return to a wild life, to come back of themselves daily to the place where they received food, and were never injured. They live very long; and it is said for a century.

The *rooks* and *jackdaws* are much smaller than the raven, and live in flocks, either in groves or in the steeples of churches, They go to a distance, in flocks, to seek their food, which varies according to the country and the season. In some places they are looked upon as mischievous, and in others as useful. We have pointed out the means for ascertaining the truth of this, in each district. In the evening, the whole flock returns to the grove or the steeple, and after uttering loud cries, go to sleep.

The *magpies*, unlike the rooks and jackdaws, live in couples in the neighbourhood of houses. Their plumage is black, with the belly and part of the wings, white. The magpie is celebrated for its cunning, and for its propensity to carry off and hide whatever it meets with. It lays up in autumn a store of dried fruits for the winter. Both sexes work at the construction of the nest. It is often built at the tops of trees, and is constructed externally of twigs plastered together with mud; and is covered by a kind of roof made of small thorny branches firmly interlaced. There is one door for entrance, and another for exit. The bottom of the nest is lined with fine and flexible roots. The female lays seven or eight eggs, on which both sexes sit alternately.

The *jay* has a more brilliant plumage than the other birds of this family; it is intelligent, and can be taught to whistle, and even to talk like a parrot.

The *blackbird*, *oriole*, and *thrush* form a small family of indigenous birds. The oriole is of a fine yellow colour, and makes a nest which is always suspended like a cradle to the fork of a branch. It fastens it in its place with grass, and also with any pieces of cord, string, or ribbon, which it can find. It is extremely rare in England, though very common in Southern Europe. The blackbird has the reputation of being cunning, and the bird-catchers have some difficulty in taking it. The male is black, with yellow beak; the female is brown above, and varied with grey and reddish brown on the throat. During the fine season, it is not uncommon to see blackbirds frequenting gardens even in the middle of towns. They eat both fruit and insects. The nest is very rapidly constructed, sometimes in less than a week, in bushes or low trees. It is made of moss and mud outside, and of dried grass inside. The female lays from four to six eggs. The young ones eat nothing but insects when they are very young; afterwards they like pulpy fruits, such as

grapes, rotten apples, or juniper berries. The thrush, is brown above, and yellowish, spotted with black beneath. It sings better than the blackbird, and is therefore more

**Head of Thrush**     valued by bird-catchers.

*Warblers.*—The family of warblers includes the warblers, the nightingales, the skylarks, the robins, the titmice, the redstarts,

and lastly the wrens, which are the smallest of our native birds. The majority feed on insects, and many eat nothing else, such as the nightingales, which must be fed, even in captivity, with worms. When the nightingale

**Wren.**     is at liberty, he sings all the time the female is sitting, as if to amuse her. Among the warblers, the reed warbler generally fastens its nest to some reeds a little above the water. The skylarks have also a joyous song, which is best to be heard when they are flying straight up into the air. The titmice are scarcely larger than the wrens. They are lively little birds, active, and courageous, and destroy a great number of insects. They attack wasps and bees, and can seize them

Titmouse.

without being stung, which would certainly kill them The titmice build pretty nests of moss at the fork of the large branches. They cover it outside with lichens, so that it cannot be distinguished from the trunk of the tree. This nest is entirely closed, only having an opening large enough to admit the finger ; it is lined inside with feathers and down, on which the female lays her eggs.

The *water wagtail* is a pretty little bird, which is always found by the side of the water. It may be known by its white belly, and by its step, always easy and elegant. It moves its tail at every step it takes, from which habit it derives its name.

*Sparrows.*—This family also includes the *buntings*, the *ortolans*, the *goldfinches*, the *canaries*, the *chaffinches*, the *grosbeaks*, the *linnets*, and the *bullfinches.* They may all be known by their short

Head of Grosbeak.

straight beak, thickened at the base, and pointed. They are great eaters of corn, and are for the most part formidable to agriculture, except during the whole period that they are making their nest, sitting, and rearing their young ; for at these times they live only on grubs and insects, and feed their brood with the same.

The *buntings* live in woods in summer, and in the winter they come in flocks into the farm yards, and settle on the dung to seek for what grain remains. The young ones leave the nest before they can fly, but these birds seem to have a great family attachment, and when the young are grown up, they often continue to live with the parents.

The ortolan like the oriole, is a very rare bird in England, though common on the continent. They live among vines and cornfields ; and are caught and fattened for table, as a delicacy.

The *canary* came originally from the Canary Islands ; but it is bred in domesticity, and is one of the commonest cage birds in Europe. It is reared for its beautiful yellow colour, and because it readily learns to sing and whistle.

The *sparrow* is well known to everybody as a bold, thievish, impudent bird, found both in town and country, and plundering the barns whenever they can get into them. The nests of sparrows are always sufficiently substantial structures, but badly made. They are built on trees or in holes. The question as to the destruction of birds chiefly concerns the sparrow, and it is certain that at seed time and harvest, and also in winter when it has the opportunity, the sparrow eats a great deal of corn. But on the other hand, it rears a numerous family in its nest, and meantime the parents do nothing but go in search of caterpillars and insects to feed their voracious brood. To prove this, it is enough to look at the ground under a sparrow's nest; it is frequently covered with the heads and wings of insects, which the birds have rejected as too hard for them. In the case of the sparrow, still more than in that of other birds, it ought to be asked if the mischief which it does at some seasons is not really more than compensated by the good done by it at a season of the year when it spends the whole day in destroying the swarms of insects which eat the germs of plants, and the buds of fruit.

Another familiar bird is the Robin Redbreast, a bold and quarrelsome bird, which frequents the neighbourhood of houses in winter, in search of food; and is a great favourite with children.

# ORDER OF GALLINACEOUS BIRDS.

The *partridge* belongs to the order of gallinaceous birds.

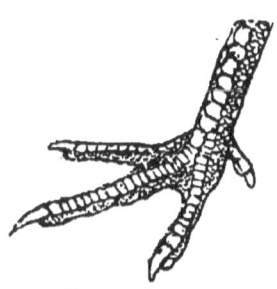

It makes a clumsy nest of dry grass in the fields, heaped up in a hole on the ground. It lays fifteen or twenty whitish grey eggs, which hatch in three weeks. The parents then show the young ones how to scratch the ground to look for ant's eggs. But as so many young ones could not be kept under the wings of a partridge, the father and mother sit side by side to protect them all.

Foot of Partridge.

The *quails* arrive here in spring, and migrate about the month of September. As they fly badly, they wait for a favourable wind to start, and only cross the seas where they can find rocks and islands upon which they rest from time to time.

The *peacock, pheasant,* and *guinea-fowl* are chiefly valued as

Pheasant.

ornamental birds. The peacock is perhaps the most beautiful of all birds; but the male alone has the well-known tail of brilliant large feathers, which can be raised and spread like a fan. The female peacock is grey, and has not this brilliant plumage. It is common in birds for the male to be more orna-mented than the female; but this difference is very obvious in the game birds; the females of the peacock, pheasant, and fowl, have less beautiful feathers than the males.

The peacocks and pheasants come originally from Asia, and

havo been brought here from thence. It is also in the East Indies only that the domestic fowl is found wild. The turkey on the contrary is found in America, and was brought to Europe after the discovery of that country.

The *guinea-fowl* was brought from Africa, as its namo implies.

The domestic fowl is of very great value for the food of man, both for its flesh and eggs, in which a large trade is carried on. There are a great many breeds of domestic fowls which have somewhat different qualities ; some are prized for the delicacy of their flesh ; others are particularly good sitters ; some lay better than the others.   But no breed combines all these qualities.

The chickens hatch after the eggs have been sat upon for three weeks.   Thirteen eggs are generally allotted to one hen to hatch.   The eggs can bo changed, added to, or even replaced by those of another species, the duck for instance ; the hen rears them very well, and is only uneasy when they go into the water, where she cannot follow them.   The hen shows extraordinary courage in defending her chickens, even against animals much stronger than herself.

*Pigeons.*—The pigeons and doves form a family, all the members of which have a great common resemblance. They have a somewhat slender beak, always a strong flight, and feel

much mutual attachment,   When wild, they build their nest indifferently on the ground, on trees, or in the rocks ; but it is always somewhat ill constructed.   There are only two eggs, on which the male and female sit in turn.   The young are born almost without feathers, and with the eyes still closed, as in cats ; the parents are greatly attached to them, and feed

Pigeons.

them by disgorging some of the food which remains in their crop.

Pigeons are reared for food and for the sake of the dung which is collected in dovecotes, and which forms an excellent manure when mixed with other substances, such as earth.

Pigeons are accustomed when taken from their dovecotes, to return through the air from a very great distance. The breeds which can thus retrace their route are called *carrier pigeons*. They have sometimes been made to undertake very long, journeys, but they always succeed much better in spring and summer, than during the winter. A cage of pigeons caught in a dovecote is brought to the distance first of fifty miles, then to a hundred, then to 200 or 300, and sometimes to 500 or 600 miles or more, and the pigeons are afterwards set at liberty. They are then seen to rise to a great height, turn round several times in the air, and then all at once take flight with a sudden start in the direction of their dovecote, where they arrive at the end of one, two, or three days, worn out with fatigue. Carrier pigeons have often been used to carry messages; and the service which they rendered during the siege of Paris, in spite of an exceedingly unfavourable season, is well known. In winter, they travel with much more difficulty, and find their dovecote much less easily than in spring or summer.

---

## ORDER OF WADING BIRDS.

The *ostrich* is at once the largest of this order, and the largest

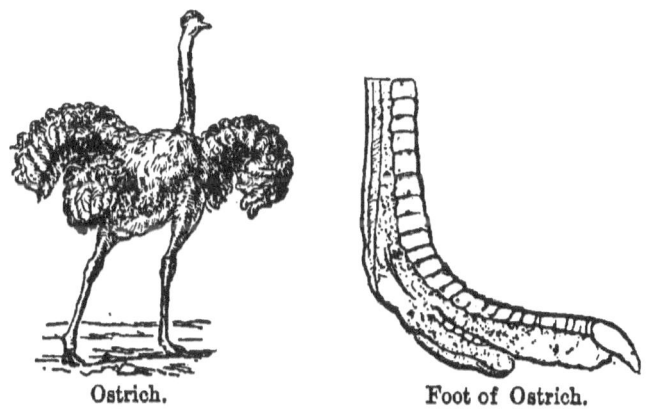

Ostrich.          Foot of Ostrich.

known bird; it is about six feet high, and its body is as large
as that of a horse. The ostrich inhabits the deserts of Africa;
it cannot fly, but runs with very great swiftness. Its eggs are
larger than the head of a child, and of a fine yellow colour;
the shell is very hard. It lays them in the sand, and the heat of
the sun is sufficient to hatch them. The ostrich is hunted for
the beautiful feathers in its wings and tail. There is a very
large trade in them.

The *cranes* are very large birds which visit our country
occasionally. They have an ashy grey plumage, and make long

journeys. Our climate
is too warm for them in
summer, and they then
fly away towards the
north; in the winter
they return towards the
south. Their flight is
strong, and they are
pre-eminently migratory

Cranes.

birds. When they are about to start, they assemble in flocks,
and arrange themselves in two files united in front, and
diverging behind. They always preserve this order, and are
seen to fly thus at a great height in the air. The bird at the
apex of the triangle only remains there for a certain time, and
then falls into the rear, or at least attempts to do so, and another
takes its place to cleave the air.

The cranes like other wading birds, put their head under
their wing when they sleep. They also often lift up one leg,
and stand for hours together motionless upon the other.

The *herons* have an ashy coloured plumage with a black crest
behind the head, and the front of the neck white, spotted
with black feathers. They disport themselves during the
day on the borders of lakes and rivers, and at night
retire to the woods, or to the groves which are reserved for
them, and which are called *heronriés*. They make their nests

G

Heron.

as high on the trees as they can, and prefer the summits of poplars. The heron has large wings; its flight is powerful, and it can soar very high. When it is pursued by a bird of prey, this is its means of escape, and it tries to rise above it. It is also extremely patient when it watches for its prey on the edge of the water, and remains there for hours without stirring.

The feathers of which head-dresses are made, are procured from a small white species of heron, called the Egret, which is found in America, and which likes to perch on the horns and back of buffaloes and oxen.

The *storks* also feed on mulluscs, which they fish for in the

Stork.

waters; but instead of being wild like the herons, they seem to like the society of man : they make long journeys like the cranes to seek for a warmer climate during the winter, and they return in spring to build their nest in houses and chimneys. They are often seen in the towns and villages of Holland and Alsatia, where they even arrange places for them when making the roofs; every house is glad to possess a nest of storks, and they take care not to do them the least injury. They are never hunted and never caught, and it is noticed that the same couple returns every year to take possession of the same nest. They are very rare in England.

# ORDER OF WEB-FOOTED BIRDS.

The *gulls* have a powerful flight; they live on the borders of the sea, and make their nests in holes on inaccessible rocks. They have fine white plumage, which makes them very conspicuous on the wing; and they feed chiefly on fish. When a storm threatens, the gulls fly restlessly backwards and forwards, uttering shrill cries which the sailors well understand. It is not rare to see them carried inland by the wind, and flying in places very far from the sea; but they hasten to return to the coast. When they are fatigued at sea, they rest on the waves; and they can swim as well as they can walk and fly. (Foot of web-footed bird, see p. 64.)

The *cormorants* are dull-coloured birds, which live like the gulls by the seaside, and feed like them on fish. The cormorants stand on a rock, and remain motionless until they perceive their prey, when they dart into the water and seize it. The cormorant can be tamed, and used to catch fish; but a collar must then be fastened tightly round his neck, and not being able to swallow the fish, he brings it back.

Cormorant.

The *pelican* is not found in Britain, though it is met with in some parts of Southern Europe. It also lives on fish, but it generally fishes in rivers; it has an enormous beak, and below

that, a great elastic pouch in which it puts the fish before swallowing it, or when it wishes to bring it to its young. The pelican has a fine white plumage; but when it returns

Head of Pelican.

G 2

to its nest with the fish that it has killed, it sometimes happens that the front of its neck and breast are spotted with blood ; and this no doubt has given rise to the fable that it pierced its breast to feed its young; but this story is no truer than a thousand other fables related of birds.

The swans, geese, and ducks, form a family of water-birds characterised by their broad and flattened beak.   They have all very downy feathers, which are largely used for making bedding.

The *swan* is reared in domesticity for the beauty of its plumage, but it is also met with in a wild state in great marshes.   It makes its nest among dry reeds, and lays seven or eight greenish grey eggs, in the month of February.   The female sits upon them for six weeks, but the male does not leave her, and defends her against any enemies who might disturb her.

The *goose* is a valuable bird in the poultry yard, but it is also found in this country in flocks in a wild state.   It makes great migrations, and flies like the cranes in a triangular arrangement, in order to cleave the air with more ease.   Geese do not deserve the reputation in which they are held ; they are intelligent animals, although they do not appear so.   In domesticity, geese afford quill-pens and down.   The former are the wing feathers which are pulled out twice a year.   They then undergo a preparation which makes them brittle, and capable of being cut with the knife.   When geese are reared for the table, they are allowed to feed at large with us ; but on the continent they are shut up, and given as much to eat as they can swallow ; and they are sometimes even put into small cages where they have scarcely room to move.   The animal then grows fat, and yields a highly valuable grease.   At the same time, the liver has grown to two or three timer its former size ; it is taken out after killing the bird, and is used to make *patés des foies gras*, for which Strasburg is especially famous.

*Ducks*, like geese, are very valuable for food, and their feathers are also useful.   On the continent they are fattened like geese, and their *foies gras* are even more highly esteemed.

The wild duck passes the summer in the north, and returns to us about the month of October. It arrives in small flocks which travel in the evening or by night, but which make an easily recognisable noise in flying. They disperse themselves among the marshes, and along the banks of rivers.

# CLASS OF REPTILES.

## DIAGRAM 5.

Reptiles are vertebrated animals; that is, they have a skeleton like mammals, birds, and fish, but their shape is very different, as may be seen by the *tortoises, lizards, serpents, frogs,* and *salamanders,* which are reptiles. They are at once distinguished from the birds and mammals in not having warm blood; they are cold. Among all animals, birds and mammals alone have. warm blood.

The bodies of many reptiles are covered with scales. They nearly all, like birds, lay eggs from which the young ones emerge; they breathe air by lungs like mammals and birds, but their respiration is very slow, and their heart does not beat so fast. The rapidity of their breathing increases a little when they are warm, and they are then sometimes very lively; but cold benumbs them, and they can scarcely move. They are generally silent animals, only uttering a rather low hissing. The frogs must be excepted, which make a loud and very disagreeable croaking.

Reptiles have been divided into four orders; the *Chelonia,* which comprises the tortoises; the *Saurians,* which are the lizards, the crocodiles, and the blindworms; the *Ophidians,* including all serpents, whether venomous or not; lastly the *Batrachians,* under which are arranged the frogs, salamanders and newts. These four words are derived from the Greek, and exactly indicate in that language the animals which represent each order.

# CHELONIA—DIAGRAM 5.

THE TORTOISES.—The tortoises are sluggish animals, which look as if they were enclosed in a cuirass. This is formed by a

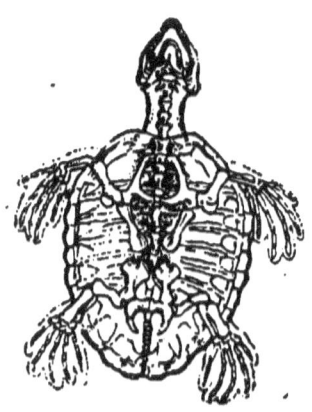

horny substance which covers bony plates. It is then quite evident that tortoises cannot go out of their carapace. When the scales upon it are torn a little, the blood runs immediately. Tortoises are also remarkable for their horny beak, which much resembles that of birds, while all other reptiles have teeth like mammals. There are marine tortoises, and land tortoises.

Tortoise.

The sea tortoises, or *turtles*, sometimes reach a very large size,

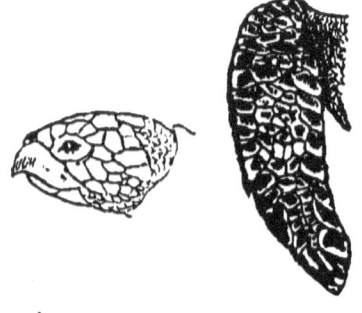

being upwards of two yards in length ; their front legs are flattened, and arranged like fins for swimming ; and they are sometimes met with at a great distance from land, floating on the water. They lay their eggs on desert and sandy coasts. They are also hunted for their flesh, which

Head of Turtle. Fore leg of Turtle.

resembles calf's head, when cooked. To catch them, they approach them by night when they are on land, and turn them on their backs ; they cannot turn themselves over again ; and are killed.

The land tortoises are not so large, and their forelegs, instead of being made for swimming, are strong, and armed with claws

with which these animals dig holes where they hibernate during
the winter season.

The carapace of the turtle is covered with large plates of a fine
brown colour. These plates form the substance known in
commerce as tortoiseshell, of which combs and many other
articles are made. The plates are thin, but they melt the
tortoiseshell and can then give it any required thickness.

## SAURIANS—DIAGRAM 5.

*Saurian* is derived from a Greek word meaning lizard.

LIZARDS.—The reptiles of this family are only represented in
this country, by some little grey and greenish lizards, which are
found along old walls, in the hottest days of summer. Their
activity is wonderful on sand or stones, where the sun falls ; but
as soon as night comes on, or it grows a little cold, they become
torpid. In spite of their small size, they are courageous, and if
you hold your finger to a lizard which puts its head out of a
hole, it darts up, and bites it with its sharp little teeth. They
usually feed on insects and slugs. But one curious peculiarity
is that when one of these reptiles is seized by the tail, the tail
remains in the hand, without the animal seeming to suffer from
this mutilation ; and when the tail has thus been broken off, it
soon grows again.

The same thing is noticed in another animal which is found in
our woods, and which is formed like a serpent, and called the
*blindworm* or *slowworm*. It also is very fragile, and breaks off its
tail when seized by the end of the body. The blindworm has
no limbs, and glides like a serpent ; but nevertheless its gold-
coloured eyes are protected by eyelids, whereas serpents have
none. It is a very gentle animal, and quite harmless, for its
teeth are too weak to hurt anyone, and it thrusts out its little

black tongue now and then, which is bifurcated at the end, like that of all lizards and serpents, but which cannot do any injury.

CROCODILES.—The crocodiles are a kind of large lizard which inhabit the rivers of hot countries. They sometimes grow to a considerable size, and attain a length of five or six yards. They have a great number of pointed teeth, and are very voracious.

Head of Crocodile.

They live chiefly on fish. They come to bask in the sun on the bank, and only move on land with difficulty; but they recover all their agility on the water, where they can dive for a very considerable time.

## OPHIDIANS—DIAGRAM 5.

*Ophidian* is derived from the Greek word *ophis*, which means serpent. There are in this country only two races of serpents, one of which, the *viper*, is venomous, and the other, the *common snake*, is not. The viper is the smaller; it may be known at once by its yellow colour, with a broad undulating black line along the back. On the top of the head, this black line is double, and forms a V.

The viper only is venomous. On opening the mouth of a dead viper, which must always be done with great caution, because there is still some danger, we find, in addition to a number of fine sharp teeth, two teeth much larger than the others. They are situated on each side of the upper jaw, close against it, and partly covered by a fold of skin. These teeth are called *fangs*, and are not firmly fixed in the jaw like the others; they lie against the gum, or are raised at the wish of the animal, by a

joint at the base.   On closely examining one of these fangs, we
see in front, towards the point, a small slit, and on breaking it,
we find that it is hollow like a tube.   This channel in the tooth,
and the slit form the passage for the poison.   This is secreted by
a gland placed in the middle of the muscle which raises the fang
when the animal wishes to bite; the muscle in contracting
presses on the poison-gland, and the venom runs into the wound
through the channel in the tooth.   The fangs being thus
moveable at their base, are not very firmly fixed ; and the viper
often leaves them in the flesh ; but they are soon replaced by
others concealed in the gum, which grow and take the place of
those torn out.   The presence of these fangs always allows us to
distinguish the bite of a viper from that of a common snake,
even before the poison has begun to work.

In a bite from a harmless snake, all the teeth make similar
holes, like large needles, but in the bite of a viper, two holes
larger than the rest are visible, which are caused by the
fangs.

The bite of a viper is always dangerous ; it will make a man
very ill, and may kill a child.   When one is bitten, the first
thing to do, as in any other accident, is to send for a doctor.
While waiting, it is always advisable to make the wound bleed
as much as possible, and to suck it, provided there is no sore on
the lips or in the mouth, through which the poison drawn from
the wound might enter.   The wound should also be washed
with alkali or ammonia, if there is any at hand.   Lastly a rather
tight bandage should be placed on the wounded limb ; above the
elbow if the wound is in the hand ; and above the knee, if it is
in the leg.   The bandage should never be drawn tight enough
to make the limb cold, stiff, or insensible.   But the doctor ought
to arrange this.

The *common snake* has no fangs, and is therefore not venomous.
It is very easily tamed.   It is very fond of milk, but it would be
quite impossible for it to suck the cows, as it was formerly
believed to do.   The common snake and viper change their skin

every year.  The epidermis loosens in a single piece, first round
the lips, and then the animal moves backwards and forwards
among the stones, to shuffle off this epidermis along the whole
length of its body, till it finally comes out of its old skin like a
glove.

Serpents feed only on living prey, which they swallow without
tearing or bruising them.  A viper swallows a mouse or a small
rat, at a single mouthful, in the following manner.  It springs
on it, kills it, and seizes it by the head.  The jaws of the
serpent are then seen to distend enormously, and by little and
little it swallows a prey larger than its own body.  After it has
swallowed it, it lies motionless for a time, as if fatigued by the
exertions which it has made, whilst its head returns to its usual
size.

There is in America a very large kind of serpent, the *boa*,
which can swallow a sheep in this manner after crushing it in
the coils of its body, or against the trunk of a tree.  The boa is
not venomous.

There are however two other kinds of serpents which are
much more venomous than the viper; one is the *spectacle snake*,
and the other the *rattlesnake*.  The spectacle snake inhabits
India.  It owes its name to a pattern on its neck which almost
exactly resembles one of those pairs of spectacles which were
formerly worn, like eyeglasses, on the nose.  These serpents
have the power of inflating their neck with air, which gives
them a peculiar appearance.  They raise themselves on their
tail when they are irritated, and in some countries the jugglers
exhibit them in public, but they take care beforehand to remove
their fangs by giving them a piece of cloth to bite, which they
jerk sharply when the animal has buried its teeth in it.  After
this, they are no longer dangerous, or at least their bite is no
more to be feared than that of the common snake.

The *rattlesnake* inhabits America, and is one of the most
venomous known.  It has a row of hard horny pieces at the
end of the tail, which make a noise when rapidly shaken; and

End of tail of rattlesnake.

it is from this peculiarity that the snake derives its name.

All serpents, like lizards, have a forked tongue which they sometimes dart out, and which is sometimes improperly called their *sting* ; but it is soft, and it is quite impossible for them to do any harm with it. One ought always to destroy as many vipers as possible in a country, but there is no occasion to destroy the common snakes. Vipers, unlike most snakes, do not lay eggs, but bring forth their young alive.

# BATRACHIANS, DIAGRAM 5.

The name of this family is derived from a Greek word meaning frog ; and it also includes the salamanders. All these animals much resemble other reptiles, but they differ from them in having no scales, but a naked skin, and especially because they come out of the egg in a different form from that which they will afterwards assume ; they thus undergo what is called a *metamor-phosis*. A frog, for instance, lays eggs. The eggs are transparent as jelly, and we soon see the vitellus (which is not yellow

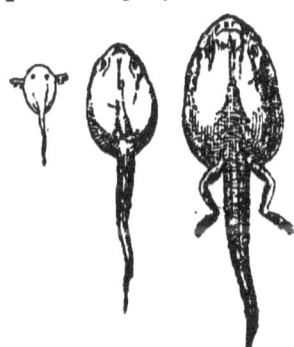

Tadpole.

as in the fowl, but brown) transformed into an animal which has no resemblance to a frog ; it is composed of a large head and a tail, and is called a *tadpole.* It has two tufts on each side which are gills, and it has no lungs. It does not breathe the air of the atmosphere. But there is always a certain quantity of air in water ; and this is what forms small bubbles on the sides of a vessel in which water is boiled. The tadpoles breathe this air by means of their tuft-like gills. Afterwards they disappear, and the tadpole grows larger, but without changing its form. It lives on water plants; and at last two legs which are of no use, but which will afterwards become the great hind legs of the frog, grow from the end of its body, on each side of the root of the tail. Presently the tail decreases, and the fore legs appear; and afterwards the tail disappears altogether, and we then see a little frog which begins to grow to its full size. But from this moment its life is completely changed. It has no longer gills, but lungs; it is obliged to breathe air, and likes to come out of the water; it lives no longer on plants, but eats insects; the frog has completed its metamorphoses. All the batrachians undergo metamorphoses more or less similar to this. In the first stage, they are said to be in the larva state. Frogs are very easily taken with a hook baited with a bit of red rag. They are not eaten in England, but the hind legs of a common continental species are considered a great delicacy in France.

Toads live on land rather than in water; they eat small slugs and insects, and are consequently useful animals in gardens, which ought not be destroyed. They come out of their hiding places on damp evenings. If anyone offers to seize them, they fill their lungs with air, and swell. At the same time they discharge their urine in order to escape more quickly, and it was thought that they projected venom, but it is no such thing, and the toad is not venomous as is generally supposed, or at least it has no venom except in the small tubercles which cover the skin of its back. But as it has no means of injecting it into the body of other animals, it is not in any way dangerous. The female lays her eggs in the water, and the young ones exactly resemble the tadpoles of frogs; they leave the water as soon as they have undergone their metamorphosis, and live in damp places.

There are other batrachians, the shape of which is much like that of lizards, for they have four legs of nearly equal length, and a tail. These are the *newts* and *salamanders*. The *newts* live in ponds, and may be known by their naked skin, and by their

Newt.

belly, which is of a fine orange colour. The males have along the back, but only in spring, a crest jagged like the teeth of a saw. The young are also born in the shape of tadpoles.

The *salamander*, which is not found in England, lives in damp places, but does not like to go into the water. It is not much larger than the newts, and may be known by its yellow marblings on the black ground colour of its skin. It is an altogether harmless animal like the newt, and we cannot tell where the fable came from that it would not burn if put into the fire.

If we wish to keep batrachians alive, it is not necessary except while they are in the tadpole state, to keep them constantly in the water. To keep frogs, for instance, the best means is to put them into a cage, or still better, under one of those covers of wire gauze which are used to preserve meat from flies. It will only be necessary to put into the cage or under the cover a saucer full of water for the frogs to bathe in sometimes. The best way to feed them is to hang a little bag of maggots in the corner of the cage or cover, and the maggots will change into flies, and the frogs will eat them.

# CLASS OF FISHES.

## DIAGRAM 5.

Fish are cold-blooded animals like reptiles, but they always live in water, and breathe by means of *gills*. They are shaped liked combs, and are of a fine red colour, as may be seen on raising the *gill-covers*. Fish breathe by taking in water by the mouth, and discharging it through the gills. On touching the gills, the air which is contained in the water, parts with its oxygen, and takes up in exchange carbonic acid from the blood, so that the respiration of fishes does not differ essentially from that of mammals and birds; only it is effected by means of the air contained in the water, instead of atmospheric air.

The limbs of fish are replaced by fins; but they swim in the water, especially when they wish to move quickly, by the motion of the tail alone. Many have a bladder in the body, which is completely closed and filled with air to assist them to float in the water. Most fish are very voracious, and swallow their prey at a single gulp.

They lay a great many eggs, but often very small ones; and when the animal is full of eggs, it is said to be full of *roe*, which is the name generally applied to fishes' eggs. It has been estimated that a salmon may lay 27,000 eggs; a pike 500,000; a sole, 100,000; a mackerel, 500,000; and a cod-fish from 3 to 9 millions of eggs.

The skeleton of fishes in not always of the same nature. In some, it is formed of hard and sharp bones : in others, on the contrary, as in the ray, sturgeon and lamprey, there are no bones, but only a skeleton composed of tough cartilages which

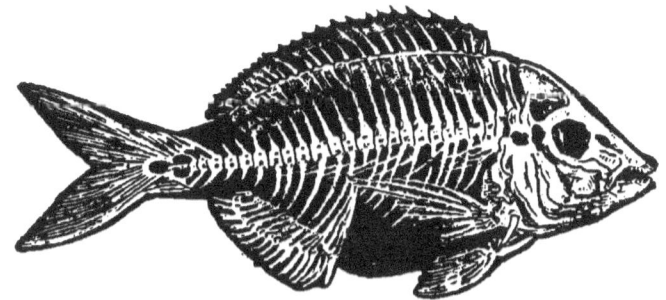

Skeleton of Fish.

break between the teeth. For this reason, fish are divided into two large orders, that of the *osseous fishes* which have bones, such as the salmon, herring, and pike ; and that of the *cartilaginous fishes*, which have none.

The habits of fish are in general very little known, and we shall chiefly speak of kinds which furnish very cheap and abundant food.

If we wish to observe the habits of various small kinds of fish, and in general of all water animals, we can always make a cheap *aquarium* with a bell glass like gardeners use, by turning it upside down between the legs of a reversed stool. In order to keep an aquarium, or rather to make it keep itself, in order, there are several precautions to be taken. We must put at the bottom some small pebbles, and a stone or two, and especially some flints with a hole in them. Care must be taken to set the aquarium in a place which is neither too dark nor too much exposed to the sun. It is good to suspend in it a flower-pot with an aquatic plant. Some water-lentils might also be strewn over ; but it is necessary that the surface should be very little covered. Lastly, it is important not to put too many animals into the aquarium, nor too large ones, nor carnivorous animals, which would eat the others. People often forget that the animals in an aquarium must be fed like a bird in a cage, or a dog which is kept chained up. We must try to find out suitable food for the animals that we rear; no doubt some will find what they like in the water; but it will not do to depend on that. If there are not many animals, and if there are some plants with them, and too much food is not thrown in so as to spoil the water, it will preserve its clearness for a long time, and it is unnecessary to change it. At the end of some time, we shall observe some small insects and molluscs which we did not know to be there. The surface of the glass will be covered with slime, but it can be cleaned without disturbing anything, with a piece of rag tied on a stick. A seawater aquarium can be easily managed when it has once been stocked. It is then enough to make a mark to show the level of the water. When it has sunk by evaporation below the mark, we must add fresh water as far as the mark, and we can in this manner keep various marine animals alive for a very long time.

*Carp.*—The carps form a large family which also includes the *barbels, tench, whitebait,* and *goldfish*; the scales of the carps are large and rounded, and these fish all feed on plants, and only rarely touch animal food.

The carp inhabits our rivers, but it also likes ponds, where it sometimes grows to a very large size. It has been proved to live a very long time, at least 150 years.

The *whitebait* is sought after for the sake of a kind of silvery dust under its scales, of which false pearls are made. It is also considered a great delicacy, for which Greenwich is especially famous.

The *goldfish* was brought from China about a century ago, and has multiplied so much in our country that it is now very common in a semi-domesticated state.

*Salmon.*—The salmon family only includes the salmon and trout. Young salmon have their skin marked with small

H

coloured spots, like trout.   Salmon live in the sea, but they
ascend the streams and rivers every year, sometimes to a long
distance, to spawn.   When they have deposited their spawn,
they return to the sea till the following year.   The flesh of the
salmon is red, and highly esteemed in countries where these
fish are not common, as in England and France.   Their eggs
are comparatively large.   They can, after being laid, be carried
to a great distance in damp moss, to be transferred to rivers
which the salmon do not generally ascend.   When the young
are hatched, they remain at first motionless at the bottom of the
water; they then begin to swim, and make their way towards
the sea.

*Herrings.*—The herring family includes the *herrings*, the
*anchovies*, and the *sardines*; and it is one of those which are most
useful for food.

The anchovy is chiefly found in the Mediterranean, though it
is not uncommon on our coasts.   It is generally eaten potted, or
made into sauce.   The sardine is common on the coast of Brit-
tany, where they fish for them with floating nets on the surface
of the water, and preserve them in two ways; the first are either
salted and put into barrels, or else they are fried, and put into
tin boxes with oil, which are soldered up.

The herring appears on our coasts in shoals like the sardines,
but it forms larger *banks* as they are called ; and it also appears
later; and while they fish for sardines in the summer, the herring only begins to appear about the end of September or the
beginning of October.   They are then taken in prodigious quantities, and cost almost nothing in seaport towns.   The boats go
out to fish for them by hundreds.   They are taken by floating
nets of great length, in which they entangle their gills.   When
there is good fishing, it does not take more than two hours for
the net to be loaded with fish.   The herring is sold fresh, but
is also preserved in various ways; it is salted, pickled in vinegar; or *smoked* by putting it into the smoke of a fire made of
resinous wood,

The *tunny* and the *mackerel.*—The tunny is a large fish which sometimes reaches the length of a yard and a half. It is not very common on our coasts, but they fish for it in the Mediterra nean, where it appears in large shoals. It much resembles the mackerel. It is taken either with lines or nets, in which a great number are killed by the blows of a boathook.

The mackerel is especially abundant in the Atlantic Ocean, and in the Channel ; and is one of the most important of our English fish ; as it is frequently taken in enormous quantities.

*The stickleback.*—This is a very little fish which lives in ponds, rivers and brooks. It may be known by the spines on its back

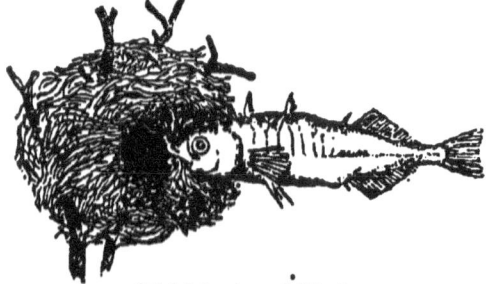

and sides. It erects them when it is threatened, and in- flicts painful wounds with them, though not venomous ones, as was formerly be- lieved. Sticklebacks

Stickleback and Nest.

are active, and assume in Spring very fine blue and red colours which they lose. They make true nests at the bottom of the water, they collect small pebbles and weeds, and lay their eggs there, which both parents watch unceasingly. They never ab- sent themselves, and keep the water in constant agitation near the nest. It is easy to observe all this in shallow pools, or in rivers overhung by trees, if they are approached carefully, and without frightening them.

*The Cod.*—Of all fish used for human food, the cod-fish is that

which is taken in greatest abundance. Every year the various European ports despatch hundreds of vessels

Cod.

to fish for cod on the American Coast. They are found on our shores, but not in sufficient abundance for a ship to be rapidly loaded with them.

They go to fish for them off the coast of Iceland, and especially near the island of Newfoundland, at a point where the sea is not very deep, and which is called the *bank of Newfoundland*. Ships arrive there by thousands from all countries to pass the fishing season. The cod is taken by lines, and its voracity is such that it is unnecessary to select the bait to put on the hook. When the fish is brought on deck, its head is cut off, and it is split open all along. The eggs or *roe* are laid aside to serve as bait to sardine fishers. The liver is used to make *cod-liver oil*, which is a valuable remedy for eruptions, scrofula, and diseases of the chest. Lastly, the cod being thus opened at the belly, is spread out, and laid between two layers of salt; after some days, this first pickle is thrown away, and the salt is renewed; and the fish thus prepared is put into barrels to be brought to Europe. Sometimes instead of salting the cod, they are content to spread it open, and dry it. It becomes as hard as a board, and is then called *stockfish*.

*Flat-fish.*—The family of flat-fish includes the *plaice*, the *dab*, the *flounder*, the *turbot*, the *brill*, and the *sole*. All these fish have a peculiar appearance which is not noticed at first. They may be known by the brown back, where the two eyes are placed and by the white belly; but one is struck by seeing their mouth awry. When they are cleaned, it is also observed that the intestines are only on one side. These fish have a very peculiar structure. In-

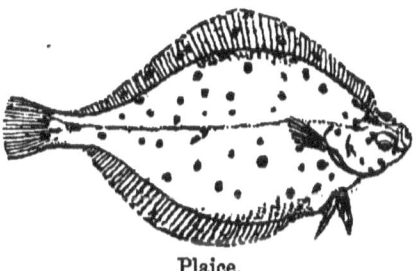

Plaice.

stead of having the belly below and the back above they have a white side turned towards the ground, and a brown side turned towards the sky; and the eye which would be of no use to them if it was below, has travelled round to the side of the eye which is uppermost. To place one of the fish of which we speak in its true position, so as to compare it with a herring, for instance, we must place it in a position which it never takes itself, the dark side to the right, and the pale side to the left.   And·then we shall see that all the parts except the eyes are in the same position as in the mackerel.   The tail is vertical ; the bones directed above and below ; the mouth is horizontal, and the gills and the intestines have their usual position.   These fish are therefore animals which live on one side, and swim on one side.   They are all very good for food. .

*The Eel.*—It inhabits the sea, the rivers, and even the smallest ditches ; it will even live in a bucket or a pan.   It can be reared thus, and will grow for years, and reach its largest size, or about a yard long.   The eel feeds on small fish, worms, and frogs.   In the spring time we see large rivers full of prodigious quantities of very small and nearly transparent eels, which make their way up the stream towards its source.   They can then be taken by thousands by merely dipping with buckets.   Eels, like several other fish, have no scales on the skin ; this is used to make thongs which are valued for their toughness.   A fish is caught in the sea which is very like the eel, the *conger eel* : it is however much larger and not so long in proportion, it sometimes grows to the thickness of the thigh.

*The Sturgeon.*—This is a large fish, the body of which is covered with plates of bone as rough as files.   Its head is prolonged in front, and beneath it is a narrow mouth, and it can only feed on small marine animals, in spite of its large size.   It lives in the sea, but it breeds in rivers, where they fish for it.   Its flesh and eggs are articles of great trade in Russia.   The eggs

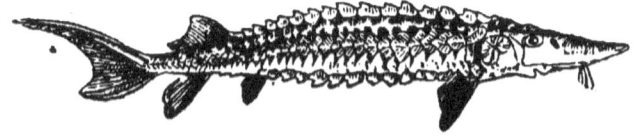

Sturgeon

are sold under the name of *caviare*.   Lastly, they make *isinglass*

of its swimming bladder, which is used in many trades. The
bladder is cut into small pieces, and dried in the sun. Isinglass
is also employed in cooking to make jellies, but it is most valuable
for manufacturing purposes. The sturgeon is very rare in Britain, and if one happens to be caught in the Thames, it becomes
a perquisite of the Queen.

*The Ray.*—The ray is a flat fish like the turbot and the plaice ;
but it is enough to look at one for a moment to perceive that in
the ray, the white side is really the belly, and the brown side the
back. The mouth, placed under the pointed head, is in its
usual place, and the intestines are really in the middle of the body.

The rays sometimes grow to a considerable size, and their
mouth is armed with pointed teeth crowded together. They live
chiefly on crabs.

The rays, instead of laying a great number of small eggs like
those of other fish, lay only a few, and these have a very peculiar

form. They are nearly square, and flattened, with the four angles prolonged
into a point. The egg is protected by
a skin which is sometimes satiny in appearance. The yolk is as large as that

Egg of ray.

of a hen's egg, and floats in a transparent
albumen. In some countries these eggs are called *sea-cushions*,
and *sea-mice* ; mice, because they are silky like the skin of a
mouse ; and cushions because they have actually very much the
appearance of a small cushion with four ribbons at the corners.

*The Torpedo.*—A fish is found on the coast of England and
France which somewhat resembles a ray, and discharges very

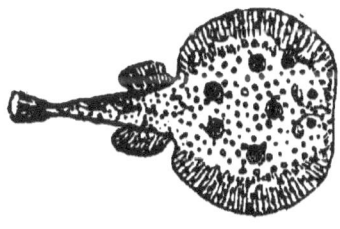

strong electric shocks when it is
molested ; this is the torpedo.
Several fish can give similar
shocks, but those of the torpedo
are the most formidable ; they
paralyse the arm, and if the animal is vigorous, the effect pro-

Torpedo.

duced by its electric discharge is

similar to that which is produced by a violent blow of a stick on the shoulder.

*The Sharks.*—These are not all so large as those which can seize the legs of a bather in the sea, and tear off the flesh.

Shark.

Much smaller fish, known as *dogfish*, are true sharks except in size. All fishes of this kind have mouths furnished, like those of the rays, with several rows of teeth, only the sharks have them very long and pointed. It is with these that they tear the prey which they cannot swallow at a gulp. At the back of the teeth which they use, others always grow, so that if by accident some are broken or lost, or worn out, they are soon replaced and the gaps filled up. These animals are always ravenous, and they are seen to follow ships to devour whatever is thrown into the sea. It is then sufficient to fasten a piece of meat to a strong hook attached to a chain, to catch them; if a cord were used, it might be cut by their teeth.

*The Lamprey.*—We shall finish the list of fishes, and of the cartilaginous fishes in particular, with the *lamprey*. It has a body

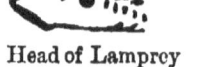

Head of Lamprey

like that of an eel, but it looks as if it had no head. It has in front only a large sucker, with which it attaches itself to rocks like a great leech.

*The Hippocampus.*—Certain fishes are so called from a Greek word meaning horse, because their head somewhat resembles that of a horse. They swim by means of a small fin on the back, and preserve the singular attitude which is represented in the figure. At other times they remain straight, with their tail rolled round some marine plant.

Hippocampus.

# CLASS OF INSECTS

---

## DIAGRAM 6.

### GENERAL OBSERVATIONS.

When we examine an insect, a cockchafer for instance, we at once see that it is very different from the animals which we have just been considering, such as the mammals, birds, reptiles, and fishes. The insects, and all the animals which we have still to notice, have no vertebræ or skeleton ; for this reason they have been called *invertebrata* ; that is, animals without vertebræ. But among the invertebrate animals there are some, like insects and crustacea, which have a shell formed of rings which are more or less hard, and jointed together. These animals have been formed into a separate sub-kingdom called *articulata* or *annulesa,* from this structure. The limbs, as we may see in the cockchafer or the crab, are also formed of small hard cylinders jointed together. The name insect itself comes from a Latin word which implies *formed of separate parts*

Insects undergo *metamorphoses*, that is, they do not emerge from the egg in the form which they will assume later. They also change their skin at different periods, which is called *moulting*. Insects generally undergo two metamorphoses ; they consequently pass through three stages ; that which lasts from the

time that they emerge from the egg to their first metamorphosis
is the *larva state;* that which separates the first metamorphosis
from the second, is the *pupa, nymph,* or *chrysalis state;* and the
third state is that of the *imago* or *perfect insect.*

*Larva state.*—In this state, insects have often the form of a

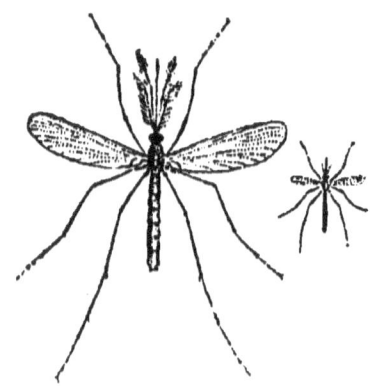

worm, and this name is applied
to them; the larvæ of butterflies are caterpillars; the
white worm will become a cockchafer; the maggot will become
a flesh fly; and thus with a number of insects.

The habits of larvæ often
differ much from those of the
perfect insect which proceeds
from them. In the first place
larvæ never fly. There are many
larvæ which live in water while the
perfect insect is aerial: this is the case,
for instance, with the *gnats.* Other
larvæ live underground, like the *white
worm,* while the cockchafer lives in
trees; larvæ do nothing but eat for

Gnat magnified.

Larva of gnat magnified.

almost all their lives, and the greater part are therefore injurious
to man. Larvæ have a very variable number of legs, sometimes
none at all, like maggots. Lastly, they have frequent moults
according to their growth.

Many larvæ spin a cocoon like the silkworm, in which they
enclose themselves to undergo their first metamorphosis.

*Pupa or chrysalis state.*—This state is that in which the silk-

worm is found in the cocoon when it
is opened before the moth emerges.
The silkworm is contracted together;
it is generally of a brown colour, and
it moves the rings of the hind part
of its body which terminates in a

Chrysalis.

point, but these are the only movements which it can make. It has no visible limbs, and does not eat. All the pupæ of insects, however, are not thus immoveable; and there are some which much resemble either the larva or the perfect insect.

*Imago state.*—After the chrysalis has remained motionless for some time, the skin which envelopes it tears or splits, and the perfect insect emerges in the form which it will henceforth retain. It is the structure of the perfect insect which we are now about to try to describe.

The body of an insect (we may take a cockchafer as an example) seems to be entirely composed of a definite number of solid rings regularly arranged, and forming three very distinct

Antenna of
cockchafer.

Antenna of
weevil.

Antenna of
moth.

regions, the head, thorax, and abdomen. These rings, as also those which form the limbs, are called segments.

The head is provided with a mouth, antennæ, and eyes. The antennæ are a kind of small horns which are found in many articulated animals. They have very different shapes, as may be seen by comparing those of a moth, a cockchafer, and a weevil. In the moth or butterfly, they are formed of a great number of very small joints placed end to end. In the cockchafer they are formed of plates which seem to form a fan. In the weevil, the antennæ are elbowed at a right angle. Insects use their antennæ to feel the objects or the ground round them.

The mouth differs much in different insects; but it never

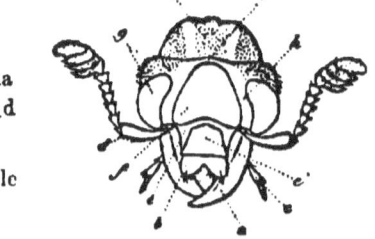

j. i. Neck
g. Eye
d. Antenna
f. Forehead
c. Palpus
d. Mandible

Mouth of insect.

resembles that of vertebrates; there is never a moveable lower jaw. The mouth of an insect can be well examined in a large green grasshopper. We shall then see that the jaws or *mandibles* as they are called, move laterally; they are situated to the right and left, and open and shut sideways, to seize or crush their food, instead of moving up and down. All the articulata which are provided with jaws have them lateral. But all insects are far from having the mouth constructed like that of a grasshopper. Many have in the place of mandibles, a *proboscis* with which they pierce the skin of men and animals; such as the gadflies, the bugs, and flies. It is enough to look at a fly resting on a piece of meat or sugar, to see that it has, instead of a mouth, a *proboscis* with which it imbibes its food. Butterflies have also a long rolled-up proboscis, which they unfold to plunge to the bottom of flowers, in order to suck up the nectar. On each side of the mouth or proboscis, insects have a second pair of antennæ much smaller, and called *palpi*. They are very well seen in the grasshopper.

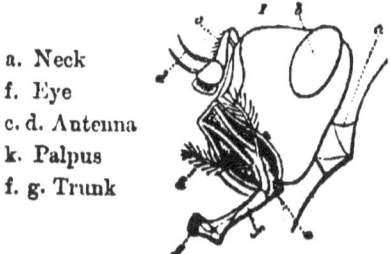

a. Neck
f. Eye
c. d. Antenna
k. Palpus
f. g. Trunk

Head of fly magnified.

The eyes of insects and other articulated animals are not formed like those of vertebrata. They have neither pupil nor crystalline lens. We see when we look at the eye of an insect, only a convex surface, looking as if polished, and with a peculiar lustre, which is sometimes reddish, as in the flies, or greenish, as in the dragonflies.   If we closely examine the eyes of a crab, we see that this convex eye has the appearance of a riddle with a quantity of little holes.   Each of these little holes is really an

eye, and all the articulata, insects as well as others, have consequently, instead of one eye on each side of the head, two clusters of eyes, each of which is too small in many cases to be distinguished without a magnifying glass.

The *thorax* bears the legs and wings.   Insects have always *six* legs.

Portion of insect's eye magnified.

Spiders have eight legs, and form a separate class; most of the crustacea have ten; others more or less; some myriapods have often a considerable number.   All insects do not fly; many, like fleas, lice, and chigoes have no wings.   Others have only two wings, as the flies.   Others again have four.   Among the last, there are some in which the four wings are alike, like butterflies, dragonflies, and wasps; but there are other four-winged insects in which the two forewings sometimes

Spider.

differ much from the hind wings, as in the cockchafers and grasshoppers.   In this case, the first pair of wings is called the *elytra*.

The thorax is chiefly filled by the muscles which move the wings and legs.   The digestive organs are in the abdomen.   Insects have special organs for respiration, which are also found in spiders, but which are neither lungs nor gills.   When the

Woodlouse.

wings of a cockchafer, for instance, are lifted up, we discover a small imprint on each segment of the abdomen, on each side. These are called *stigmata* and are the openings for fine vessels filled with air, which communicate with the exterior by the stigmata. These vessels are called *tracheæ*. They have a fine silvery colour, and nothing is easier than to see them by opening the abdomen of a cockchafer under water, with a little care. All these channels, which look like threads of silver, and which ramify in the middle of the organs of the insect's body become visible immediately. It breathes by means of these tracheæ.

Insects lay eggs, which are often very numerous. They take care to deposit them in places where the larvæ which will emerge from them will find the means of living. If the larva is aquatic, the insect though aerial, deposits its eggs in the water.

Insects, chiefly in consequence of the great appetite of their larvæ,

· Eggs of bug on a leaf

are injurious animals; they are the scourge of agriculture, and man suffers much more from them than from tigers, lions, or venomous serpents. There are certainly some insects which eat others, and are consequently useful auxiliaries to man; but there are not very many. And it is on account of the devastation of insects that all the insectivorous mammals and birds ought to be considered as the greatest friends of agriculturists. Some insects however must be mentioned, which, like the bee, the silkworm, and the cochineal, are directly useful to man, but these are exceptions.

Insects are divided into several orders, which are characterised either by the structure of the mouth, or by the number and texture of their wings. We shall mention each of them, and state their distinguishing characters.

# ORDER COLEOPTERA.

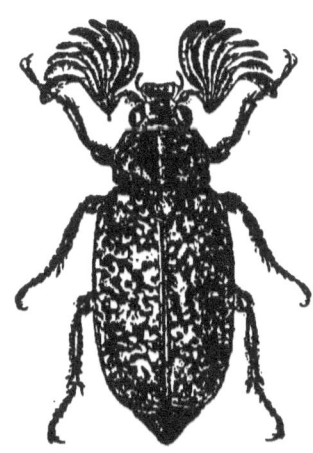

The insects which belong to the order Coleoptera have four wings. The two first are hard and horny, and are called the *elytra*; the two others are thin, transparent and membranous; they close them by folding them transversely, and then place them under the first wings, which cover them like a sheath. This can be seen very well in a cockchafer which has just been stopped in its flight; the wings are still unfolded, and reaching beyond the elytra, under which they are gradually seen to disappear

Great cockchafer, male.

Coleoptera undergo a perfect metamorphosis. They have all lateral jaws for bruising. They can in some cases bite with them, but they do not pierce like gnats. They have no poison apparatus at the end of their abdomen like bees. A beetle can therefore always be safely taken into the hand to examine it, for it can do no injury beyond nipping the skin, sometimes a little roughly, with its mandibles.

*Cicindelidæ.*—These are small carnivorous beetles, and consequently useful to man. The cicindelidæ or tiger beetles are known by having their corslet (which is that portion of the thorax between the neck and the base of the wings) narrower than the head and elytra. These are insects which fly in full sunshine. They are of a beautiful green, with yellow spots. They are swift, and pursue their prey, which always consists of small insects, with great eagerness. In the larva state, the cicindelidæ dig a cylindrical hole in the ground, carrying away the earth and gravel. Their head has a hollow above which they use for a hod; they go and empty it away from the hole,

Cicindelida.

resting from time to time when they are too heavily laden. 'When their pit is completed, they wait at the entrance with their head out, and watch for ants and small insects. The cicindelidæ prefer dry and sandy ground.

The *Brachiri, or Bombadier beetles* much resemble the cicindelidæ; they derive their name from the power they possess of making a small detonation with the end of their abdomen when pursued, and discharging a disagreable vapour which stops or drives away the pursuing enemy.

The *Carbai* are beetles which do not fly. They have elytra, but on lifting them up, we find no wings underneath. Their corslet is square. The carabi are very active : some are black, and others have fine metallic colours of green or gold. They destroy great numbers of caterpillars, and so they are the greatest friends to man, of their class. Their body exhales a strong odour, and when seized, they discharge a blackish, fetid, acrid liquid from the mouth. If it falls into the eye, it produces a very acute pain, which is however not followed by any serious consequences.

Carabus.

The *Calosomæ* feed only on caterpillars. They live in trees, and their voracity is extraordinary ; even when they are satiated, they run after caterpillars, bite them with their mandibles till the viscera extrude, and then leave them to die ; they hunt their prey unceasingly, and render valuable service by destroying many very destructive caterpillars.

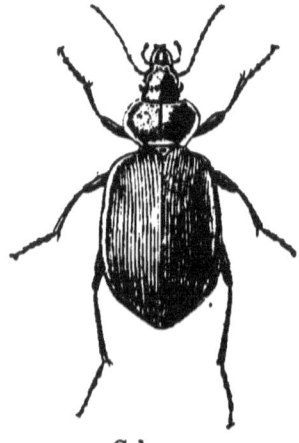

Calosoma,

The *Dytiscidæ* swim, and pass part of their life in the water. For this purpose, they have flattened legs, shaped like oars. But they are obliged to come to the surface frequently in order to breathe the air, a store of which they always keep under their elytra, when diving to the bottom of the water. They are essentially carnivorous animals, and even attack newts, and devour them alive. In the evening, they leave the water, spread their wings, and fly about. They then fly into rooms, attracted by the light. When touched, they exude from the surface of their body an oily liquid, as white as milk, and extremely fetid. The larvæ live constantly in water, and are carnivorous like the perfect insect. They have pointed mandibles which cross each other, with which they can pinch severely.

The *Gyrinidæ* are small insects which are seen twisting round like drops of quicksilver on the surface of ponds, where it is very difficult to catch them. They are popularly called *whirligig beetles*. Their back is black, but is so highly polished as to reflect the light of the sun like a metal button.

The *Hydrophilidæ* somewhat resemble the Dytiscidæ in form and habits, but are much larger. These are our largest native water-insects. Like the Dytiscidæ they pass the day in ponds. They swim and fly very well, but walk with difficulty. They can remain under water for a long time, but are nevertheless obliged to come to the surface from time to time to breathe. In the evening, they fly about. Their larva are very large, black and wrinkled, and swim with ease. They live exclusively in the water. They are extremely voracious. They are also remarkable for their habit of making themselves soft and flaccid, as if dead, when

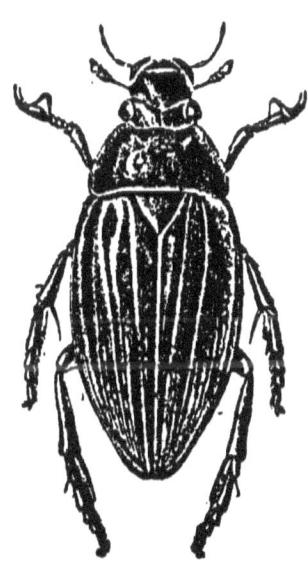

Hydrophilus, male.

seized. When the time comes for them to undergo their first metamorphosis. they leave the water, and dig a hole in the ground on the bank, closed on all sides. There the larva changes into a pupa, and this into a perfect insect.

The *Staplylinidæ* may be known by their square elytra much shorter than the abdomen, the segments of which extend beyond the wings. Their antennæ are inserted in front of the eyes. They run quickly and fly easily. When they alight, they immediately hide their wings under their elytra ; but as the latter are very short, the wings must be folded three or four times to fit under them. When threatened by danger, they raise the end of their abdomen straight up. There are many kinds of staplylinidæ, most of which are very small ; the largest and commonest is popularly called the Devil's Coach-horse ; it is quite black, and is found running on pathways. They are all very voracious, and eat either insects or carrion, they also eat one another. The staplylinidæ are generally to be met with in damp places, especially under stones.

In looking for insects it should be remembered that one of the best methods of finding them is to turn up as many stones as possible, and the largest which can be found ; for a great number of insects always select such hiding places, and are sure to be met with there.

The *Buprestidæ* have elytra which cover the whole abdomen, and have also antennæ serrated like saws. One common species is a small insect of a beautiful bronzy green. The larva lives in wood ; the perfect insect lives on trees and flowers ; the Buprestidæ are very like the click-beetles, but do not jump like them. They fly swiftly. When seized, they contract their legs and *sham death*. They remain thus for a very long time motionless, and it is only by little and little that they begin to move one leg, then two, and then they fly away very quickly as soon as they believe that the danger is over.

The *Elateridæ*, or *Click-beetles* also sham death like the Buprestidæ when an attempt is made to seize them, and let themselves

fall to the ground. If they fall on the belly, they soon stretch
out their legs to escape ; but if they fall on the back, they are
then seen to unbend themselves like a spring, and jump to a
great height. If we take an elater, it is enough to lay it on a
table, legs uppermost, to see how it acts. It first becomes rigid
and raises itself on its head and tail. All at once it unbends
itself; the corslet and the base of the elytra strike the tables
and the rebound throws the insect into the air to a height of
several inches. If it falls on the legs, it runs away; if it falls
on the back, it begins again. As it makes a slight noise on
unbending itself, it has been called the click-beetle. If held in
the hand, it attempts to perform the same manœuvres ; and every
time it unbends, it emits a green liquid. The larvæ of the click-
beetles are found under stones, in the ground. They are called
*wire-worms*, and often do great mischief by gnawing the roots of
corn.

The *glowworm*, which is found by the side of roads on warm
summer evenings, has no wings, and much resembles a larva. It
is however the perfect state of the female insect. The males fly
like other coleoptera ; they have four wings, two of which are
elytra. It is only necessary to put a few glowworms on a tuft of
grass at the window in the country to attract the males to fly
round them. On closely examining a glowworm, we may
ascertain that it is the interior of its body which is luminous, and
not the surface. On irritating the animal, it extinguishes this
light, and renews it when it is quiet. The glowworm is not the
only animal which emits light; in hot countries there are other
insects which produce a much stronger light. Some *fire-flies*,
as the luminous elateridæ are called in America, enclosed in a
small cage made for them, are sufficient to light a room.

The *Necrophori* or *burying beetles* are a little smaller than the
cockchafer, which they resemble. They have black elytra
with yellow transverse bands. These insects owe their name
to their habits. When about to deposit their eggs, they seek
for some small dead animal, such as a rat, a mole, or a mouse,

or else a bird, or a frog. They lay their eggs in it, and then undertake the great work of burying the carcass, at which several always assist; for this purpose, they burrow underneath it and hollow out the earth, and throw it aside, so that the body sinks little by little, and finally lies in a hole large enough to contain it; then they cover it with the earth which they had removed, and leave it. This work sometimes requires two days, although the beetles who have undertaken it, prosecute it with great

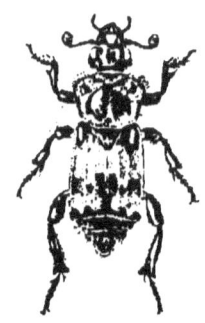

Necrophores.

ardour. The larvæ are thus born in the midst of their appropriate food. They are greyish white worms.

The *Dermestes* are small beetles about a quarter of an inch long; they have black elytræ with a white spot on each. The larva is wholly covered with hairs; it eats cheese, lard, furs, linen, and feathers. It is extremely voracious, and will even attack old bones. It is the greatest enemy of collectors of natural history, because it eats stuffed skins.

The *cockchafer* is certainly one of the most destructive insects known; in the perfect state, it devours the forests; and in the larva state it eats the roots of the crops. The larvæ are called *white worms*, and are hatched about six weeks after the eggs

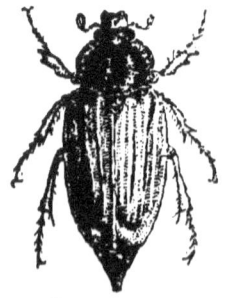

Cockchafer.

Larva, or white worm.

have been laid by the cockchafers on the ground. They

immediately burrow into it, to pass the three years before their
metamorphosis.   They burrow near the surface, and eat all the
roots which they can find, which causes the plants to die.  ·When
it begins to get cold, the grubs bury themselves deeply in the
ground, and become torpid till the following spring.   As soon
as spring comes, they again mount nearer to the surface, and
recommence their ravages for another whole year.

It is then that the grubs feel the time of their metamorphosis
approaching.   They again bury themselves in the ground,
deeper than the first time, sometimes nearly a yard.   There they
excavate a small ovoid chamber with very smooth sides.   Then
their metamorphosis takes place; the grub becomes a soft
whitish pupa, on which the limbs of the perfect insect can
already be discerned.   This pupa gradually becomes tougher,
and turns brown.   It remains thus for the whole winter.

When the month of February arrives, the second metamorphosis takes place ; the pupa becomes a cockchafer, but it is
then soft and yellowish ; and it is only gradually that it becomes
hard and acquires its colour.   About the month of March or
April, according to the warmth of the season, the cockchafer
approaches the surface of the ground, from whence it emerges at
the beginning of May, when there are already leaves on the
trees ; and then begins a new series of ravages ; it ascends the
trees, on which it sleeps during the daytime ; but in the evening
it flies about, and begins to devour the leaves.   In less than a
fortnight, the cockchafers have sometimes been known to strip
entire forests of all their foliage.   Then the female lays her eggs ;
for this purpose she leaves the trees, and with her legs which
are dentated, she digs a small trench, and lays her eggs there.
She lays from fifty to eighty eggs of a light yellow colour.
Then she dies; and three years afterwards the cockchafers
sprung from her eggs will appear, and it is on this account that
the cockchafers generally appear in greatest numbers at regular
intervals of three years.

Every means should be sought to deliver us from this terrible

enemy; but as it hides under ground, it is always difficult to get
at it. The best plan is to collect very carefully all the
cockchafers, which are found turned up by the plough; but it
must be remarked that this plan is only available when the
cockchafers are still near the surface, in spring, or the beginning
of autumn. If we wish to destroy them in quantity, we must
do the work at the exact depth where the grubs are. If they
are only two or three inches from the surface, as sometimes
happens, deep digging would turn up very few; but if they are
far from it, a superficial examination is altogether useless. But
it is easy to ascertain first with a spade at what depth the grubs
are, and consequently at what depth it is necessary to work.

In the perfect state there is only one way of destroying the
cockchafers, namely, to collect as many as possible, for which
the local authorities ought to pay as much as they can. One
remark must however be made. The cockchafers that the
collectors should be paid for, ought to be all alive, or else it is
useless to collect them; and they ought to be paid for very dear
during the first few days of their appearance, and the price
should be lowered afterwards. The reason is that during the
first few days they have not yet deposited their eggs; and are
consequently of great importance; whereas it is not much use to
collect cockchafers at the end of the season which have already
laid 50 or 60 eggs in the ground, which will produce as many
grubs in succeeding years; at this time it is useless to continue
the pursuit of cockchafers, and it is wasting public money to pay
for them.

The cockchafers thus collected form when mixed with earth a
good manure. The best way of killing them is to plunge the
bags in which they are brought into boiling water.

*Cantharides.*—The name of cantharides is often applied to all
bright green beetles. But the true Cantharis is a rarity in the
South of England, and is too scarce to be of any commercial
importance. It is met with in May and June on jasmines,
ashtrees, and lilacs. There is a great trade in cantharides for

medical purposes. They are largely collected on the Continent, where the persons employed cover the face and hands, and go in the morning and shake the trees which the insects frequent, over cloths. They are then killed by dipping them in vinegar, or by putting them in a sieve under which vinegar is boiled. Then they are put into tight fitting cases, that the mites may not get at them. But if mites can eat the cantharides with impunity, men cannot, for they are a terrible poison. They are used to make blister paste; and if we look closely at this preparation, it is easy to discover small brilliant green atoms in it, which are fragments of the elytra of the insects. These are the cantharides, which make blister paste act like a hot iron, or like boiling water, in raising the epidermis and forming a blister full of water.

The *corn-weevil*, which is also called simply the *weevil*, lives in

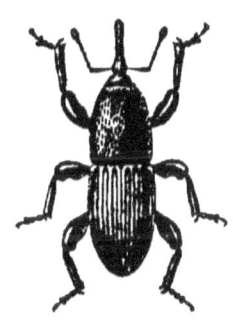

heaps of corn, keeping itself hidden near the surface without burying itself deeper than a few inches, and without ever appearing outside. Its colour is a maroon brown; its corslet is covered with small points, and its elytra with very fine furrows. The mischief which the weevil may cause may be imagined when it is remembered that it lays each

Weevil.        of its eggs in a separate grain of corn.

For this purpose it drills an almost imperceptible hole. The larva which is thus born in the very middle of its food, devours the grain without coming out of it; then undergoes its metamorphosis; and it is only then that the weevil pierces the outer shell of the corn to go and lay its eggs in its turn.

It was absolutely necessary that every means of protection against such an enemy should be sought for, and it was soon observed that the weevil could only live in the middle of grains of corn which were not agitated. Quiet is absolutely essential to their development. It makes its escape when disturbed. The means of getting rid of it was consequently discovered: which is to move the corn either with the shovel, or by arranging the corn in the granaries in such a manner that everything moves whenever a sack is taken away.

Bostrichus, magnified.

The *Bostrichi* have ovoid, elongated bodies, their antennæ are short, and terminate in a club. They are very small beetles, the larvæ of which are extremely destructive to trees. They establish themselves between the wood and the bark, and then hollow out tortuous channels, which are almost always filled with the sawdust formed by their work. They live thus two years, after which they construct a cocoon formed of sawdust fastened together by filaments of silk. They pass the winter there in the pupa state, and it is only in the following spring that they emerge from their prison in the perfect state.

Larva of Bostrichus, natural size, and magnified.

The *Longicornes* are a family of Coleoptera which may be

Longicorn.

Larva of Longicorn.

known by their slender antennæ, often recurved like a ram's horns. The larvæ of the longicornes also burrow long galleries in wood. In order to escape, they sometimes pierce the sheets of lead with which the timbers of roofs are covered. They cut this metal with their mandibles, as easily as the wood itself.

*Coccinella.*—We will end the list of beetles by a small insect known to everybody as the *lady-bird,* or *lady-cow.* It is the coccinella. It is carnivorous and destroys the plant lice; and is therefore a friend of man. Unfortunately its small size does not allow it to give us great assistance. But it seems that we do not misunderstand this friend, for everyone avoids hurting it.

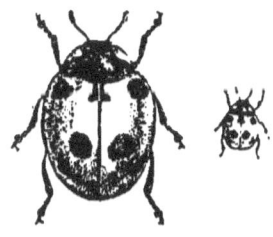

Coccinella.

# ORDER LEPIDOPTERA

This includes the insects commonly called *butterflies* and *moths.* They have four wings alike, covered with a dust which adheres to the fingers when they are rubbed. Instead of jaws, they have a rolled up proboscis. Their metamorphoses are complete. Butterflies cannot do much harm in the perfect state with their proboscis, but their larvæ or caterpillars have jaws like those of Coleoptera, and often commit great ravages.

Red Admiral.

*Silkworms.*—The *silkworm,* one of the most useful insects

should be mentioned first among the Lepidoptera. It is largely reared on the Continent, but is only reared as a curiosity in England. Its life is composed of seven stages, five passed in the larva state, and the two last as pupa and imago. The five first stages occupy twenty-four days, and the two last require sixteen days. It is only after forty days that the egg, when hatched, produces a moth which lays eggs again, after which it dies. Great numbers of silkworms are reared in the south of France, in establishments provided for the purpose.

When they observe that the eggs are about to hatch, they lay mulberry leaves near them. The little worms, just out of the egg, and quite black, crawl upon them and begin to eat. They are placed on large trays, and are watched and tended with the greatest care.

Each stage ends with a moult. This is preceded by a day's sickness, during which the worm raises its head, and neither eats nor moves. When it has shed its skin, it still remains a considerable time without eating. On the other hand, its appetite is insatiable between the moults.

The first stage lasts four days; and during this period it is necessary to cut the leaves fine which are given to the silkworms, because they only eat them at the edges; but this precaution is unnecessary afterwards. The fifth stage is longest, and that in which the appetite of the worm is at first most voracious; but it soon ceases to eat; it seems to become more transparent; it tries to climb, and spins ends of silk here and there; this is called its *time of change*. Then they lay small branches of birch or broom on the trays, on which the worms can easily spin their cocoons. When only a few silkworms are reared, it is enough to put them in screws of paper, which seems to suit them very well.

The worm takes three days to spin its cocoon, after which it becomes transformed into a chrysalis. The silk issues from small openings placed near the mouth. The worm fastens its thread to some object, then draws back its head, the thread winds off, and it attaches it somewhere else. The whole cocoon

except the outer floss silk, is generally formed of one single
uninterrupted thread. It may measure 300 yards in length
If the body of a silkworm is opened when it is about to spin, we
can easily perceive two long sacs, folded on themselves. They
are yellow, and are filled with a sticky substance, which is
nothing else than silk. It becomes solid, and winds off, as it
issues into the air. In some countries, these sacs are taken from
the body of the silkworm, unfolded, and allowed to dry. They
thus obtain strips of a true silk, which are much thicker than
ordinary silk, but these strips are always very short, and are
chiefly used for fishing lines.

When the cocoons are allowed to remain in the branches
where the worm has spun them, the moth emerges at the end of
about a fortnight, pushing aside and breaking the threads. It
cannot fly, though if reared in the open air for a few generations,
it recovers the power of flight; but its legs are furnished with
hooks with which it clings to objects. It eats nothing, and soon
begins to lay its eggs. It is placed on sheets of paper or
cardboard, where it lays its eggs near each other, and then dies.
These eggs are the provision for the following year, and are
called *seed* in commerce. Great quantities are annually sold and
bought in all countries which produce silk, and they go to fetch
them from the most distant countries, where the silk is finer than
in Europe.

But the moth in emerging from the cocoon, spoils the silk and
breaks it. To prevent this, only a certain number of cocoons
are allowed to produce moths, and the chrysalides of the others
are killed by exposing them to heat. The cocoon can then be
reeled off entire, like a ball of silk, just as it has been spun
entire from a single end. For this purpose, it is put into boiling
water, after having rid it of the floss silk which surrounds it; it
is rubbed with a soft brush to find the end, and when this is
found, there is nothing more to do than draw it gently off by
means of a winder, which generally winds off a great number of
cocoons at once. Several of these threads are put together to

make the silk with which they make stuffs prized for their beauty as well as for their durability, for silk is stronger than canvas or linen.

*Bombyces.*—The silkworm belongs to the family of Bombyces, which are mostly injurious animals, because many of them eat the leaves of trees like the silkworm, without producing a valuable substance like silk, which fully compensates for the value of the mulberry leaves. Others feed on grass ; and that of the Emperor moth, one of the largest and handsomest English moths of this family, feeds on heath. The larva is green, with tufts of hair, and transverse rows of pink spots ; and the perfect insect has a large ocellated spot on each wing.

One of the most destructive insects of this family is the  *gold-tail moth,* so called from the yellow tuft of down at the end of the abdomen of the female, which she employs to cover her eggs for the purpose of protecting them from the weather. The caterpillars are gregarious, and in some seasons

Gold-tail moth.

strip the hedges of their leaves. They also form a large web, as a shelter for the whole community, who retire within it at night. To avoid any injury which might happen to the structure from the growth of the plant within it, they take care to gnaw off all the buds within their habitation, and thus check any such inconvenience.

The *Noctuæ, Tineæ, Pyrales,* and *Tortrices,* are moths of moderate or small size which fly only in the evening, or at night, and the larvæ of which are very destructive in spite of their small size. Those of the Tineæ live on stuffs ; and after having cut some wool from the cloth, they make it into cases which they drag about with them like a dress. The strips of wool are then joined together by an extremely fine silk that these caterpillars spin. As they grow, it is quite necessary for them to enlarge t heir case; and for this purpose, they rip it up, and add a new piece. If they have changed the cloth, if, for instance, they lived first on black cloth, and are now living on red cloth, the piece added to their case will be red, and the rest black.

There is another insect belonging to the same family, the *Hypenomeuta* or *small ermine moth*, which is sometimes as destructive to our hedges as the gold-tail moth of which we have spoken already. We sometimes see the hedges in summer almost stripped of their leaves, covered with a slender whitish web, and swarming with little whitish or greyish moths, covered with black dots. These are the moths produced by the caterpillars which have caused the mischief. They are also to be met with feeding on apple-trees and other plants, especially the spindle-tree, which is liable to the attacks of several species.

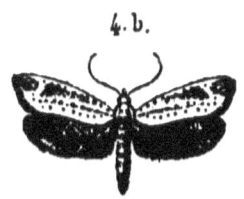

4. b.

Small ermine moth.

The larvæ of the Noctuæ live on the leaves of various plants. Some few live just beneath the surface of the ground, and do great harm by devouring the roots of plants. The fore-wings of the moths of this group are generally of some shade of grey or brown, and the hind wings are paler.

The *Sphinges* or Hawk Moths are large moths which only fly at dusk, and are therefore sometimes called *crepuscular Lepidoptera*. In the evening, the hawk-moths come buzzing over flowers, and extract the honey from them with their proboscis. One of the largest has a yellow pattern on its corslet with two black spots, which have a slight resemblance to a skull. It is therefore called the Death's Head Hawk-moth, but the resemblance is certainly not very great, and we must look carefully for it to notice any at all. This species is rarely noticed on the wing; it is remarkable for its power of squeaking, and for its habit of sometimes entering beehives to steal the honey. Its larva is as thick and long as the finger, and is often met with in potatoe-fields.

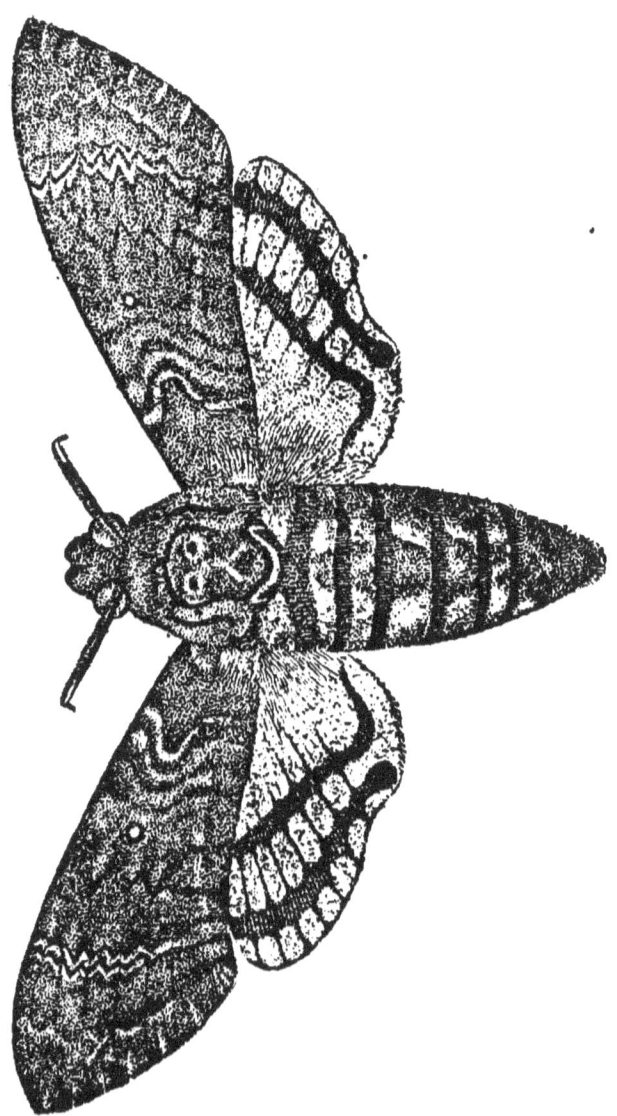

Death's Head Hawk-moth.

The  *Pieridæ*  or *Cabbage  Butterflies* have a slender body, and large wings; nearly all are white, with black spots, and feed on

various kinds of cabbage. They are found in abundance in
kitchen gardens,
which they fre-
quently devas-
tate. But it is
enough to kill
them and to keep
them away to
place branches of
flowering broom
on the plants
which are infested
by them. The insects cannot bear the neighbourhood of this
plant.

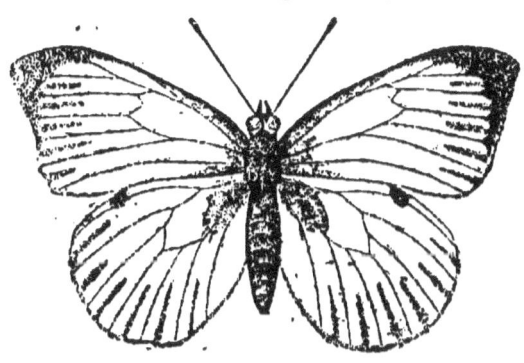

Cabbage Butterfly.

# ORDER HEMIPTERA.

Insects of the order Hemiptera have a straight proboscis
which they can easily bury in hard substances; they have four
wings, which appear at the first glance to be alike; but when
we examine them more closely, we perceive that the fore-wings
are only partly similar to the hind wings; another part, either
the edge, or a portion near the base is horny like the elytra of
beetles. Hemiptera undergo only an incomplete metamorphosis ;
the larva already greatly resembles the perfect insect, and the
pupa still more.

The *cicada* is a large insect which is found in the New Forest.
It is mute ; but most of the foreign species are celebrated for
the loud chirping of the males. The organs of song in the
cicada, are situated under the abdomen, or level with the first
segments. Two plates or two scales are noticed covering a small

Cicada.

skin stretched like that of a drum.   The cicada produces its song by agitating this with special muscles. The cicada feeds by piercing the tender bark of trees, and drinking the sap by means of its sucker.

The *thrips* has only wings in the males ; the females are wingless. They are small insects with a long body. The corn thrips causes great mischief by sucking the newly formed  grains of corn, which it does not kills but prevents from reaching their proper size.

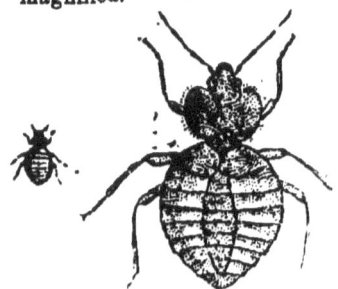

Thrips highly magnified.

The *bugs* are a family of insects which have nearly all a bad odour.   They live either on trees or in houses.   Bed-bugs are remarkable because they have no traces of wings, but they run actively. They shun the light and at night feed on blood by burying their sucker in the skin. The *Notonectæ* are aquatic hemiptera which swim by means of their long hind legs.   One half of their elytra is hard and horny; and these insects resemble tree-bugs in the form and pattern of their corslet.

Bug, natural size and magnified.

. Notonecta.

*Plant-lice.* The plant-lice are small insects which have two

projecting tubes at the end of the abdomen. The females have no wings; the males alone fly, and they are seen with their large transparent wings among the clusters of plant-lice on the stalks of the rose bush. The females move slowly, and pump

Plant-louse, highly magnified

up the sap of plants with their proboscis. They bring forth small living plant-lice, which grow round them and increase the swarms for the greater part of the year. It is only at the end of summer that they lay eggs which do not hatch till spring, and which produce males as well as females. The summer broods generally consist of females only, which are able to continue the race without the intervention of the males, until the cold weather sets in.

The *cochineal* is an insect of this order found in Mexico, from which a beautiful colour used in the arts under the name of carmine is obtained. The male has wings, as. in the plant-lice,

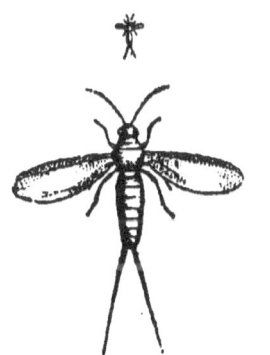

and the female has none; the latter lives on a kind of cactus, where she fixes herself, by inserting her proboscis. These plants are cultivated for the sake of the insect. When the time comes to gather them, the females are collected, and killed by putting them into boiling water, or by exposing them to the heat of an oven. Cochineal as met with

Female Cochineal natural size and magnified.

Male Cochineal natural size and magnified.

in commerce, presents the appearance of small violet grains which look like seeds, but on examining them carefully their animal nature is easy to be perceived, especially if they are allowed to soften a little in water.

## ORDER ORTHOPTERA.

Insects of the order Orthoptera have jaws like Coleoptera; they have two kinds of wings, but the elytra are soft, and the hind wings are folded like a fan, instead of being folded transversely, like those of Coleoptera. Their metamorphoses are incomplete, as in the order Hemiptera. The Orthoptera having jaws cannot suck blood like some of the Hemiptera; but they are often very formidable to the crops.

The *forficulæ* or *earwigs*, are Orthoptera furnished with a kind of pincers at the end of the abdomen, which they open with a menacing air when irritated, but with which they are incapable of doing any injury. They live in society, and are very destructive to flowers and fruit, but never get into anyone's ear, as is vulgarly imagined. The internal auditory canal is furnished in man, with stiff hairs and an acrid substance which is generally sufficient to prevent any insect from getting into it.

Earwig.

The *mole-cricket* is so called, partly because it resembles the other crickets, which are Orthoptera, and partly on account of the shape of its fore legs, which have some resemblance to those of the mole,.

Mole-cricket,

K

and which it uses in exactly the same manner. With these legs, the strength of which may be felt by holding a mole-cricket in the hand, they dig galleries under the soil, and eat all the roots with which they meet; and sometimes produce great ravages in this manner.

The *house-cricket* lives in houses near the fire-places; and the field-cricket in dry, grassy places. They are nocturnal animals, and their music is chiefly to be heard in the evening. They produce it by passing their long hind legs over their élytra, which then vibrate like the cords of a violin. Crickets and grasshoppers produce their music in a similar manner.

The *locusts*, and the *large green grasshopper*, are the largest British Orthoptera. The latter is not uncommon in England, and the former, though not indigenous, are frequently met with in some years in different parts of the country, having flown over from the Continent, and in some cases, it is supposed even from Africa. Like the crickets, their hind legs are very large, and fitted for jumping. The mischief which they cause in

Blue-winged grasshopper.

Britain is happily unimportant; but in many countries of Asia and Africa, they frequently appear in immense swarms. Sometimes vast swarms of several miles in length and breadth appear

suddenly in a country, driven by the wind. The swarms are so
thick as to form a real cloud, which darkens the sky. In a day
or two, all the vegetation in the country is devoured ; the ground
is covered with these insects; and the crops are filled with them ;
and even should they be driven out to sea and drowned, as
sometimes happens, their dead bodies are frequently washed
ashore in such quantities as to pollute the air for miles around.
When a swarm of locusts descends on a country in this manner
before the harvest, they always cause a famine. These insects
are eaten in many countries, and it is even said that certain
tribes rejoice at their approach, because they are a staple article
of food with them.

# ORDER NEUROPTERA.

The Neuroptera have four wings, which are alike in texture
and leave no dust on the fingers when touched. The nervures of
their wings are arranged in a kind of network. This is well
seen in the dragonflies which fly round ponds. The Neuroptera
have jaws, and undergo a complete metamorphosis. They have
no sting, like bees, at the end of the abdomen.

The *Libellulæ* or *dragon-flies* have a long cylindrical body, and
fly with ease. Their larvæ and pupæ live in water. They
abound in ponds and pools. They are carnivorous, and very
voracious, and attack other insects, or larger prey, such as
tadpoles. These larvæ are greyish, or of a dirty white, and
almost transparent. At the period of its metamorphosis, the
pupa, which much resembles the larva, comes out of the water,
climbs on the stalk of a reed, and fixes itself there with the
hocks of its legs. Then the skin splits, and the dragon fly
emerges. The empty skin is often found on water-plants. The
dragon-flies eat small insects and like their larvæ, devour

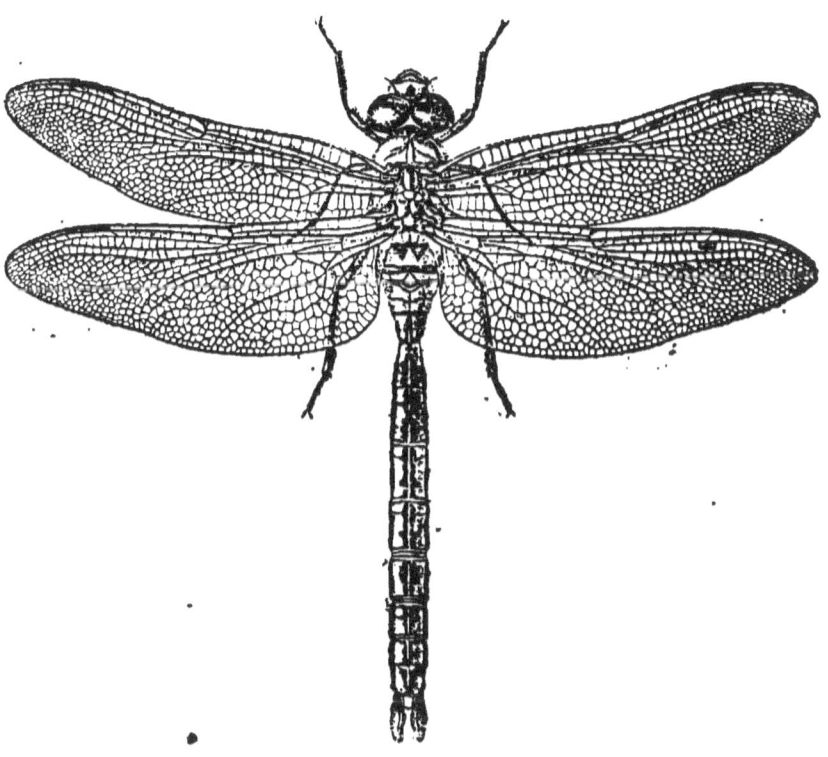

Libellula or dragon-fly.

a great number; and it is therefore to be considered a useful animal.

The *ant-lions.*—These interesting insects are not found in England, but are not uncommon on the Continent. If we happen to visit the Continent, we may sometimes notice in some places, small funnel-shaped hollows about an inch deep, and an inch and a half broad. If we look at them carefully we shall see at the bottom something like the ends of a pair of pincers, which shows us that there is an animal there. Nothing stirs however, but if we continue to watch these funnels for some time, we shall presently see an ant or some other small insect slipping to the edge, upon the treacherous sand. The ant attempts to escape; but the animal in the middle of the trap immediately throws sand over it, and makes it roll to the bottom, where it is grasped by the pincers and killed. This is not all. When the industrious insect has sucked its victim dry, it · throws it out of

the hole which it never quits itself, just as it throw up the sand before. This animal, which exhibits such curious habits, is the larva of the *ant-lion* ; and is so named on account of the great destruction which it makes among the ants. In the perfect state, it is very much like a dragon-fly, but may be

**Larva of ant-lion.** distinguished from them by its antennæ terminating in a knob.

The habits of the *termites* or *white ants*, resemble those of the ants; but they inhabit tropical countries. They have fortunately

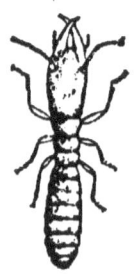

Neuter termite.

Soldier termite                                      Female,

not yet been introduced into England, but they have succeeded in establishing themselves in some of the French ports, where they commit great ravages ; they devour books, registers, wainscoting, and all descriptions of wood-work. These animals are obliged to avoid the light, and their work is always concealed. They devour a beam, but they only enter it by the ends which are fixed in the walls ; they do not make a single hole throughout its whole length ; and nothing is visible externally, until the house falls, when the beam is found to be completely eaten away inside.

The *phryganeæ* and *ephemeræ*.—We must mention two Neuropterous insects which are well known for their interesting

habits. We meet with larvæ in ponds which drag after them a case about three quarters of an inch long. When they walk about with their house, they put out their head and legs, and when they are alarmed, they hide in it. These are Phryganeæ or _caddis-flies_ in the larva state. These cases are formed of the materials which the insect finds near it; perhaps small stones, or shells, or else small twigs or pieces of leaves which the larva cuts for itself. All these objects are joined with threads of fine silk. The phryganeæ are always very curious objects to observe in an aquarium, and they can be made to construct beautiful cases by pulling a larva out of its case, and putting it into a vessel with glass beads, when it will use them to construct a new house, if it can find nothing more suitable.

The ephemeræ are remarkable for only living a day at most; but this is only true of the perfect insect. Before reaching it, the ephemera has lived in the larva and pupa state for one, two, for three year. These larvæ are generally found in rivers. They are small, and may be recognised by three slender filaments at the end of their abdomen. They swim by jerks. At last the time for their metamorphosis arrives. It is then that the ephemera lives really very quickly. The pupa emerges from the water, and is transformed about the period of sunset; it flies to some distance, changes its skin with the same rapidity; lays its eggs; and when the night has become quite dark, all the ephemeræ which emerged from the water at sunset are already dead. They consequently live less than an hour in the perfect state, after having lived two or three years in the larva state.

---

# ORDER HYMENOPTERA.

The order Hymenoptera includes the two insects which

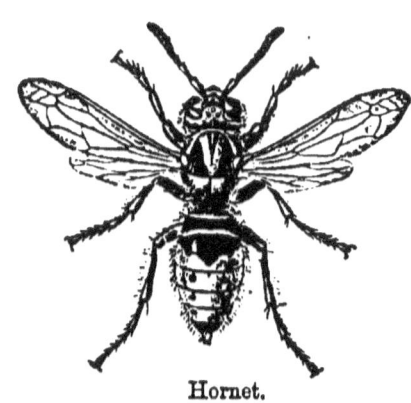

Hornet.

exhibit the most remarkable habits ; the bee and the ant. It is characterised by the possession of four wings of· similar structure, but the nervures of which are longitudinal, instead of being finely interlaced, like the wings of Neuroptera. Their metamorphoses are complete. The females have often a sting at the extremity of the abdomen, in the perfect state.

The *Tenthredinidæ* or *sawflies* have a prolongation at the end of the abdomen, or rather, a borer, dentated like a saw, from

Tenthredo.

which they derive their name. The female bores holes in the bark of plants with this instrument, and deposits an egg in each. The larvæ which emerge from them eat the leaves and buds of fruit trees, and injure them very much.

The *bees* live in society, and work in common. A hive is never inhabited by more than a single society. This is always composed of several males, one female, and a great number of bees, called *workers*, or *neuters*, which are undeveloped and sterile females. We find a similar arrangement in the societies of ants.

The *males*, or *drones*, are larger than the workers; their thorax is hairy, and their legs are not adapted for working, like those of the workers. They consequently perform no labour.

The *female* or *queen* has a very long abdomen. Her legs are not adapted for working, and she has no other occupation than laying eggs. There is never more than one queen in a hive.

The *workers* may be known by their small size. Their hind legs have a very remarkable structure. One part is triangular, hollowed out above. On examining the bees which enter and leave the hive, we discover that they bring home part of their spoil in these hollows, as if in *baskets*. The next part of the leg is equally remarkable ; it is square, and provided with several rows of short, rough hairs, which make it look like a *brush*, which is the use made of it by the insect. We often see the bees dive into flowers, and come out covered with the pollen, which is yellow for instance, in the lily, and black in the tulip. The bee is quite covered with this dust. Then it stands still for a moment ; it brushes itself with the square part of its legs, and carefully removes what it finds on its body ; it gathers it into its baskets, and goes on to collect more from other flowers. We shall see presently what it does with this pollen.

The workers have a sting at the end of the abdomen, the puncture of which is rendered more painful by a venom which is simultaneously injected into the wound. To see the sting well, it is enough to push a bee against a pane of glass with a straw. We then perceive, after several trials, its dart, which is scarcely a line in length ; and several small drops of venom at the end, as clear as spring-water. Bee stings are not generally dangerous to man ; but they can make a child very ill. When a strange animal enters a hive to eat the honey, the workers immediately rush upon it, and pierce it with their stings till it dies.

Man rears bees for the sake of their wax and honey. The wax forms honey comb. The construction of the combs is the great occupation of bees.

Sting of bee magnified. It can be observed by making them work under a bell glass covered by a basket hive ; it is enough to remove this to follow all the details of their life and labours. Bees make the wax themselves. When we take hold of one, we see that the segments of the abdomen overlap, and partly cover each other. In each space we find a slender flake of wax which gathers there, and is secreted by the skin, like perspiration with us. The bees remove these flakes with their legs, and form *combs* of them, by building and moulding the wax with their mandibles. Each comb is composed of two rows of openings, or *cells* connected at their base. These cells are always very regular in shape; and they have six sides separating them from the six surrounding cells. All the bees work together to build the comb, which increases gradually at the edges; it is always vertical, so that the cells are hollowed horizontally on its two faces. The provision brought home by the bees on their legs is therefore not used to construct combs, for it is not wax, but is employed for another purpose. The workers make a paste of it with which they stop up any holes which may exist in the hive; they plaster up the places where it does not stand even on the plank, so as to keep out draughts, and leave no opening beyond that required for an entrance. When this entrance is larger than they like, they reduce its size with the same plaster, which is called *propolis*. It is also used for another purpose; if a large caterpillar, or butterfly, as sometimes happens, penetrates into the hive; and they cannot throw it out after killing it, they cover it with propolis, and make a kind of tomb over it, which prevents them from being inconvenienced by the putrefaction of the corpse.

Bees do not make honey, but simply collect it from flowers; for it is the sugared nectar which these contain. They take as much as possible and swallow it, but it is not digested; and on reaching the hive, they disgorge it, either to feed the larvæ, of which they have the charge, or to store it up in the cells, for food during the winter, when the flower season is over. All the honey which we use is only the winter provision of the bees which we appropriate to ourselves. As honey is only the nectar of flowers, it is better in proportion as the flowers in the neighbourhood of the hive are more odoriferous; whence it follows that there are different qualities of honey.

Part of the cells are destined to contain honey for the winter, and part to contain the eggs, and rear the young larvæ. When the queen is laying, she deposits an egg in each cell. She is always followed by several workers, who see that all goes right. If the queen has accidentally left two eggs in the same cell, the workers take one out, and place it in an adjoining empty cell. After the eggs are laid, the workers never cease to tend first the eggs, and then the larvæ, which form what is called the brood.

The larva grows and changes its skin several times within the space of six or seven days; when it ceases to eat, and is about to undergo its first metamorphosis, the workers close the opening of the cell with a wax covering which the young bee gnaws its way through when it has reached the perfect state.

When a new generation of bees is thus born, there is no longer room enough in the hive; but the young ones still remain there as long as there is no new queen. But as soon as a new queen has emerged from the pupa, all the new generation goes with her to look for a dwelling elsewhere; and this forms what is called a *swarm*. All the bees of the swarm fly together, and afterwards assemble in a compact mass on some tree in the neighbourhood. They can then be taken all together, and put into a hive, where they soon begin to work, and make combs in their turn.

In order to collect the honey and wax, the bees are driven away or stupified. Then the combs are removed. In some countries, where the bees' wax is very fine, it is eaten with the honey; in other countries, the honey and wax are collected separately, by melting the latter. Wax thus obtained is yellow, and it is whitened by different methods, and is then called *virgin-wax* in commerce. Bees' wax is not used in the manufacture of sealing-wax, which is made of vegetable resins.

*Wasps* and *humble-bees* also make combs of more or less regularity, which are found in woods, with the cells partly filled with honey, and partly with brood. Wasps' combs are not formed of wax, but of a substance resembling grey or brownish paper. We sometimes find pretty wasps' nests as large as walnuts which only contain a small number of cells. These nests are made by the female only, who works, instead of doing nothing, like the queen bee, and she alone rears the larvæ which emerge from the eggs which she lays in the nest. But these larvæ produce workers which immediately begin to build the large nests which are found in hollow trees, holes, and sometimes under the roofs of houses.

*Ants.*—Ants are not less interesting than bees, although they are of no use to man. Ants, like bees, live in colonies, consisting of *males, females,* and *workers.* Only there are a great many females in each anthill. The workers have no wings, and are the ants which are seen everywhere, and which are always so interesting to observe.

. Ants have no wax, and build underground dwellings which sometimes contain a great number of halls and galleries, extending a long distance below the surface. To watch an anthill and its neighbourhood is one of the most interesting spectacles that can be imagined.

In the morning before sunrise all is quiet in the neighbourhood, and the anthill is closed, no opening is visible for the exit of the ants, which are all inside. But when the sun has risen, we begin to see some workers which clear away the soil, and make doors by which other ants soon come out. In the evening, these gates are closed, and they are thus opened every morning, and shut every night.

However, the other ants go in all directions, along paths under the grass and moss, which correspond to their highways, and sometimes extend very far. They come and go, meet, stop, and touch each other's antennæ as if to speak. Those which return are generally loaded. Sometimes they have much difficulty in carrying their burden, which may be a twig, a bit of dead leaf, a cockchafer's wing, or even a whole cockchafer. But then several ants join in carrying so large a body ; some pull, others push, and at last it arrives at the gate of the ant-hill.

If it is too narrow, other workers enlarge the gate to allow the body of the cockchafer to pass, and then reduce it to its former dimensions.

The workers are entrusted with all the labours; they build and repair the house, they take care of the eggs, and feed the 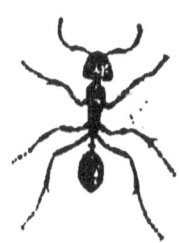 larvæ. They also seek the honey of flowers, with which they feed the white and almost motionless larvæ. Then if the weather is fine, and not too hot or too cold, as these larvæ cannot walk, they bring them into the sun, and lay them on the anthill. When a little rain falls, or they are disturbed, they remove these larvæ, which are larger than themselves, with their mandibles, and descend into the interior. These larvæ are generally called *ant's eggs*. If we look at them, we shall see the segments of their body, and perceive that they are not eggs; the eggs themselves are much smaller, and the ants only bring the larvæ to the sun when they are already large.

Worker ant.

The larvæ spin a cocoon in which they undergo their metamorphosis, and when the time comes for them to emerge, the other ants assist them, and afterwards arrange all the old empty cocoons in a corner of the anthill.

Ants are generally courageous. When they are molested, or an attempt is made to destroy their house, they rush out in great agitation, and some try to drive away the enemy while others repair the damage. When fighting, the ants rise on their hind legs, and bring their abdomen forward, from which they discharge small drops of a transparent and very acrid liquid at the enemy.

It often happens that an anthill is inhabited by two kinds of workers, one large, and armed with strong mandibles, and the others smaller. But it soon appears that the two kinds of workers seem to play a different part in the community. The large ones, those armed with strong mandibles, do not work. They seem to rest all day, but if an enemy threatens the anthill, they come out to attack it: and these are especially soldiers; the others are specially labourers, and take care of and repair the dwelling; they fetch provisions, and even bring food to the ants who do not take the trouble to go in search of it. These working ants are not the same species, but the captives of the others, who sally forth from time to time to plunder the nests of feebler species to obtain workers. The slave-making ants are not natives of Britain, but are found on the Continent.

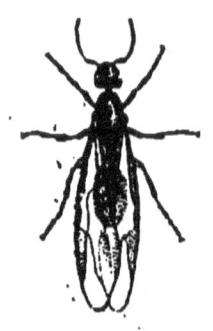

**Female ant.**

The males and females of ants have large wings with which they rise in the air, and we sometimes see considerable swarms carried about by the wind.

The habits of ants differ much according to the country which they inhabit, and according to the situation of the anthill, and it is always very instructive to observe and study those of our native species. They do not all make their dwellings in the ground. Some prefer trunks of dead trees, in which they sometimes hollow out very long galleries, with rooms at various intervals.

The *ichneumon.*—Sometimes when we crush a caterpillar in spring, we observe that its body is full of other living larvæ. These larvæ are those of a small hymenopterous insect called an ichneumon-fly, which has its abdomen furnished with a very long point or ovipositor. The ichneumon is a winged insect, and when the female is about to lay, she settles on a caterpillar, pierces its skin with her ovipositor,

**Ichneumon.**

and deposits her eggs in the body of the caterpillar. The eggs hatch, and the young larvæ feed on the flesh of the caterpillar, which suffers a living death while these larvæ devour it. At last it dies, when the larvæ of the ichneumon emerge, and spin round its body the cocoons in which they undergo their metamorphosis.

The *Cynipidæ* are other small Hymenoptera in which the ovipositor instead of being straight, as in the Ichneumons, is spirally rolled. The ichneumons are serviceable to man by destroying caterpillars; and the cynipidæ are also useful animals,

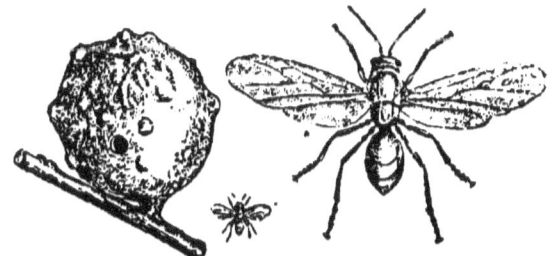

Gall nut.    Cynips of    Cynips magnified.
             the ink-gall.

but in a different way. They pierce the leaves of trees to lay their eggs there. Where they have pierced the plant, a *gall* is produced. When it is carefully opened, we always find a number of the larvæ of the cynips in the middle. In some countries a great trade is carried on in these galls, because they are used for making black ink. Those which are sold under the name of *gall-nuts* are produced by a cynips which pierces the leaves of a kind of oak.

# ORDER DIPTERA.

This may be known immediately by the insects which compose it having but two wings; they have also no jaws but only a proboscis or a sucker.

The *gnats* and *mosquitoes* live near water, and lay their eggs on its surface; the eggs are joined together in regularly shaped masses, and resemble a little boat floating on the water. The larvæ are aquatic, and swim by jerks by doubling themselves up. They rise to the surface with their head downwards. They breathe by the extremity of the body, where the trachææ open. The pupæ are also aquatic, and swim by jerks, but may be

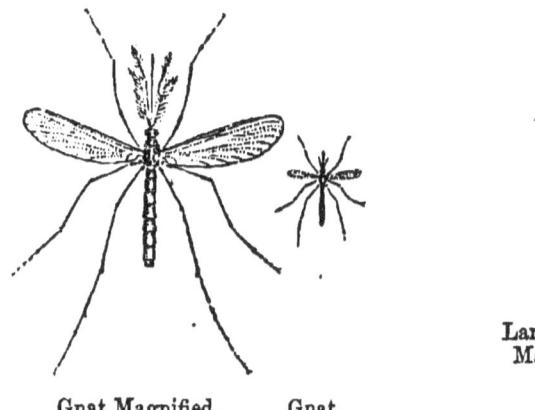

Larva of Gnat
Magnified

Gnat Magnified.     Gnat.

known by a kind of horns near the head. In their perfect state, the gnats fly chiefly in the evening, and light upon men and animals to suck their blood; they bury their sucker rapidly under the skin, and the wound which they make remains painful for some time, on account of the venom discharged into it. The shrill noise which gnats, and especially mosquitoes make when flying, is caused by the rapid motion of their wings, which strike the air several thousands of times in a minute.

The *fly*.—There are a great many different species of flies, some of which have a fine blue or green metallic colour. In the larva state the fly is called a *maggot*. Anglers use them to bait their hooks. The blue flesh-fly lays its eggs on meat which is beginning to decompose. The eggs are white, rather large, and form in small clusters. The maggots which proceed from them feed on the putrefying meat. When the time has come for them to undergo their metamorphosis, they hide in dark and dry places. When they have found a place that suits them, they shrink up, turn brown, and thus become chrysalides without making a cocoon, or changing their skin. At the end of about eleven days, the fly emerges from the chrysalis.

The old skin of the larva splits at one end, as if it could not contain the animal within it. The fly when ready to emerge, swells its head, which makes the chrysalis split; it swells itself again to get rid of the covering which surrounds it; and at last it stands on its legs; but it is pale in colour, its wings are soft, and it cannot fly, and remains quiet in one place. But in a few minutes its wings have dried, its skin has become dark, and it flies away.

The *œstrus* is a fly the larva of which has very singular habits.

The fly lays its eggs on living animals, horses, oxen, or sheep. As soon as the larvœ are hatched, they bury themselves under the skin, where they remain, and sometimes

Œstrus.      Larva of œstrus. form a small tumour in which the larvœ are found when they are opened; they only leave it when they are about to go into the pupa state, when they fall on the ground, and there undergo their last metamorphosis.

But some œstri lay their eggs on the fore legs of the horse, where he can reach them with his tongue. When he begins to feel a little pain there, he licks the spot, and swallows the young larvœ. Their strong outer skin cannot be digested by the gastric juice, and therefore the larva does not die. It attaches itself to the surface of the stomach, and continues to live there until the time for its first metamorphosis, when it drops into the food, and is expelled from the body, when it becomes a fly at the end of a certain time.

The *gadfly*, instead of having a proboscis like the flies has a sucker like the gnats. It pierces horses to feed on their blood, but always makes a large wound from which it flows in abundance. The gadfly also attacks man; its puncture is not dangerous, but it may communicate either to animals or to man

Gadfly.

a serious disease called *carbuncle*. It generally begins by a very bright red patch, in the middle of which is a black spot. When this disease is supposed to exist, the doctor should immediately be consulted. Gadflies and other insects which similarly pierce the skin may thus convey carbuncle, but it may also be contracted by touching the fresh skins of animals which have it. The flies themselves only convey it when they have previously rested on animals which have carbuncle, or have died of this disease.

# ORDER PARASITA.

All those animals are called parasites which live on or in other animals. There are also parasitical plants. All parasitic animals do not belong to the class of insects although some do. The œstri of which we were speaking, which grow either under the skin of cattle, or in the stomachs of horses, are parasites, but only in the larva state. In the insects of the order of which we now speak, some are parasites during all their life, as the louse; and others are so only in the adult state, like the flea. None of them have wings.

The flea has a sucker, with which it makes painful punctures, like the bug and the gnat. In the larva state, they are very active maggots, but these larvæ do not live on man; they live in straw, dust, and old furniture.

The *louse* attaches its eggs, or *nits* to the hair; they have a small lid which the larva raises to emerge from it. It already

resembles the perfect insect. Lice are troublesome animals, like all parasites, and are easily removed by cleanliness.

Many animals have lice and fleas which differ a little from those of man, but which can also live on the human body, and which may be got rid of by similar means. We have given a figure of the louse which infests the duck.

Nits or eggs of the louse, natural size & magnified.

# CLASS ARACHNIDA.

The class Arachnida, which at first sight seems greatly to resemble that of insects, is nevertheless very easily distinguished from it.

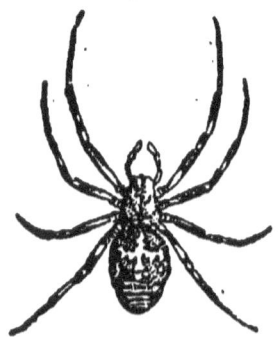

While insects have six legs, the spiders and all the animals included with them in this class have *eight* legs. There are also other differences. The head and thorax are united, in such a manner that there is no neck. The abdomen itself is sometimes joined to the head and thorax, as in the mites; and the body then forms only a single ovoid mass with eight legs. The eyes are consequently placed on the corslet; there are often 6 or 8 or even

Spider, back view.

12, separated from each other, instead of being united in a cluster like those of insects. The remainder of their structure

Side view of Spider.

much resembles that of insects; but they have not metamorphoses but only *moults*.

Many people are often much terrified at spiders, without

L 2

exactly knowing why.    They cannot generally do much harm, but
several spiders have curved hooks at the mouth, which are
furnished, like the teeth of the viper, with a venomous
apparatus.    However, our largest species cannot cause very
dangerous wounds, but it is not the same in hot countries, where
they are found as large as a mouse.    The spider is one of those animals the habits of which are very interesting to those who like to
observe them closely, and nothing is easier.    To seize their prey,
they spin nets, but they do not produce their silk by the
mouth as in insects.    It is by the extremity of the abdomen
that the spider spins.    When it is about to spin its web
anywhere, it experiences great difficulty in attaching the first

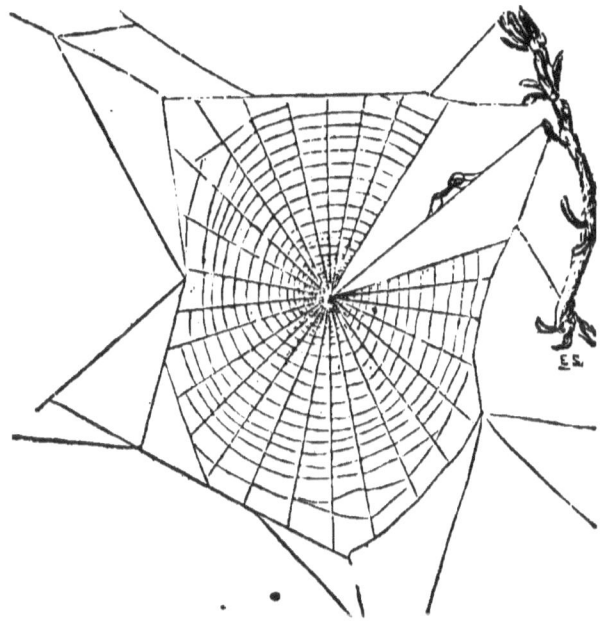

Spider's web.

threads.    Then the work proceeds quickly enough, and when
it is not interfered with, the web is perfectly regular.    It often
makes a shelter near it in addition, where it watches for the
prey which may become entangled.    As soon as a fly has touched
the web, the spider darts out, seizes it, and sucks its juices ; then

throws the corpse away, or sometimes surrounds it with silken threads and leaves it where it is. It repairs the meshes which have been broken, and returns to wait for another victim. All spiders do not lead this sedentary existence ; some species do not make a web, and simply stretch threads here and there. They are found running in the fields, and some of them leap with great agility. They are called *running* and *hunting* spiders.

Spiders are sometimes found which carry a large silken ball attached to their abdomen which they never abandon. This ball is filled with eggs which the female carries everywhere with her. Other species also put their eggs into a a silken bag, but they hang it in some part of their web where it will be secure.

It was formerly pretended that there was a spider in Italy called *tarantula*, the bite of which causes a desire to dance ; but this is a fable, like so many other tales which are told of animals.

*Mites and the itch-insect.*—The mites which live in cheese have also eight legs, and consequently belong to the class Arachnida. Their history would not be very interesting if their form did not resemble that of another animal which is a parasite on man, and which produces the disease called the *itch*. It is still smaller than the cheese-mite, and makes galleries of about a quarter of an inch in length, under the epidermis.˙ As it works chiefly at night, it is then that it causes the most violent itching. The itch is caught by contact

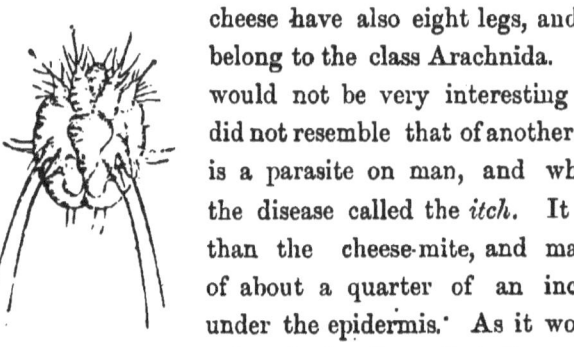

Itch-mite highly magnified.

with an infected person, when the itch-mite passes from one person to another. It is always desirable to get rid of these parasites as soon as possible ; but as they are very small and hidden under the skin, medical assistance is necessary. The dog, cat, horse, and dromedary have parasites which cause itch

in them also.    These are not the same species, but can nevertheless live under our skin, and animals can therefore communicate
the itch to man, in some cases.

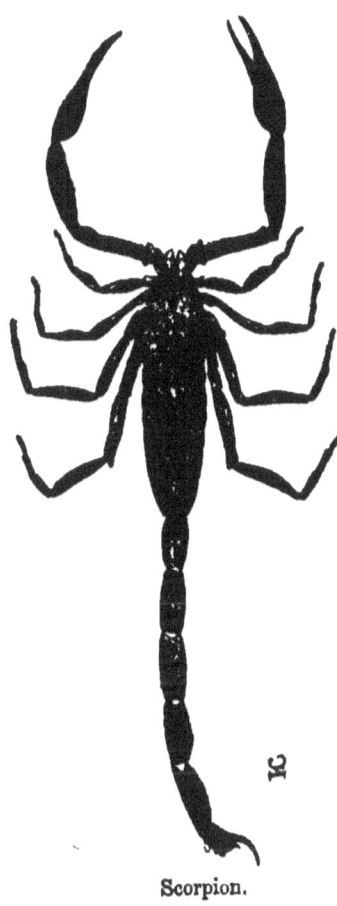

Scorpions have eight legs like
spiders, but do not spin ; their
abdomen is composed of several
jointed rings, and is furnished
with a recurved hook with which
they can pierce the skin of an
animal, injecting into the wound
a venom which causes great pain.
These animals are not found in
England, though some small and
comparatively harmless species
are common in South Europe.
They live under stones, and
under the bark of trees, and
generally come out at night.
They only use their dart when
irritated.

Scorpion.

# CLASS MYRIAPODA.

The best known animals of this small class are the *Scolopendræ,* or *Centipedes.* Their body is formed of a great number of rings,

Scolopendra.

which are all alike, except the first and last. Each of these rings is furnished with a pair of legs, and the animal advances with the aid of all its legs, which are sometimes a hundred or even more in number. Our small native centipedes live under stones, and are almost innocuous ; but the large centipedes of hot countries inflict very painful and dangerous wounds with two hooks placed on each side of their head.

# CLASS CRUSTACEA.

The crustacea, with the insects, arachnida, and myriapoda, form the great division of articulate animals, that is animals formed of segments. The crustacea may be easily known by always having more than four pairs of legs; they have ten, as in the cray-fish and crab, or a much greater number ; the wood-louse, for instance, has

Crab.

fourteen. Crustacea, like insects, have lateral jaws, and eyes in which the facets may easily be seen, but they never breathe by tracheæ : they have branchia like fish. They undergo frequent moults.

*The Crayfish.* In the crayfish the head and thorax are united.

There are five pairs of legs, but the first are much larger than the others, and resemble strong pincers. The abdomen is composed of several rings, below which are a kind of limbs called *false legs.* It is to these that the female attaches her eggs, which she carries everywhere with her. On raising the edges of the hard part, or carapace, which covers the head and thorax, we find on each side, five branchiæ branching like trees.

The crayfish which appear to be so well enclosed in their hard skins, nevertheless change them every year. When the time of moulting arrives, the carapace detaches itself from the first segment of the abdomen, and splits in the middle at the same time; and the crayfish giving a violent jerk, gets out of its old skin with an entirely new one. This is not yet as hard as the other, but quite soft; the animal then hides itself in some hole; until at the end of a few days, its new skin becomes hard, when it resumes its former mode of life. The crayfish, like most crustacea, is carnivorous.

The *lobsters* are nearly of the same shape as the crayfish, but they are much larger, reaching a foot or more in length, and ten or twelve pounds in weight; they live in the sea. During life, they are of a bluish black colour.

The *crabs* are perhaps the most abundant of all crustacea. The sea-shore swarms with them. They feed on all the dead animals and carcasses which are cast up on the beach. They are found at low tide under stones. Some are very active, and run away sideways. The common edible crab is very large, and has been known to weigh as much as twelve pounds. But they move little and conceal themselves in the crevices of rocks, and under seaweed.

Shrimps and prawns are very small crustacea which are much valued for the table, and swim with great agility in the water. They fish for them at low tide on the sand, with a net with very narrow meshes.

All crustacea similar to these which we have mentioned turn red when cooked, but very few are of this colour while living.

The *woodlice*, which live in damp places, are also crustacea;

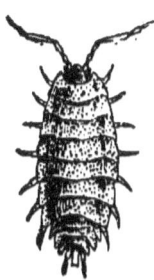

and there are some species which live in the sea. Most of these are parasitic on fishes and cetacea; and others are found among rocks and seaweed at low tide. These never grow to a large size, and have always fourteen legs.

Woodlouse.

# CLASS ANNELIDA.

The class Annelida is composed of soft animals formed of numerous segments, which have no limbs. The leech and earthworm are annelides. These animals are remarkable for their red blood. If we look at the back of a small earthworm, we see a red filament under the skin. On carefully observing it, we notice that it is sometimes broader and sometimes narrower, and that the slight dilation which it momentarily exhibits, passes along through the whole length of the animal, from back to front. This vessel, which exhibits these pulsations, is the heart.

The *earthworm*, like other annelides has no limbs, but if we look closely at a large one, and especially when we take it in the hand, we see that a great number of short stiff hairs issue from its belly, with which it crawls. It is for this reason that it is so difficult to pull an earthworm out of its hole, when it is partly in it, as it takes hold of the sides of its dwelling with this kind of grapnels. A little in front of the middle of its body, the earthworm has a swelling, which is called the *corslet*. If an earthworm is divided behind the corslet, the anterior portion does not die, and reproduces a tail; the worm becomes exactly similar to what it was before. Worms feed by swallowing earth. It must not however be supposed that this would nourish it; but there is always a large quantity of animal and vegetable refuse in the ground; it is these matters which are digested, and which form the true nourishment of the worm.   It would soon die if it were obliged to swallow pure sand.

The *leeches* are annelides which have suckers at both extremities of their body to fix themselves ; in the middle of the front sucker is the mouth.   Leeches live in stagnant water.   They were formerly very much used in medicine ; but are now employed much less frequently.   They live only on blood.   Old horses are driven into the marshes where they breed, and the leeches attach themselves to their legs.   Their mouth is furnished with three hard projections, bearing small sawlike teeth.   The leech makes the wound from which it draws blood by making these three projections act on the spot where it is fixed by its sucker.   The scars which succeed these wounds are therefore always triangular.   The leech gorges itself with blood till it can scarcely move ; but it requires a considerable time to digest this large quantity of food.   It can be made to vomit up by laying on it a pinch of salt or tobacco, or even sugar, which is a poison to some animals.

# CLASS OF INTESTINAL WORMS.

Men and animals nourish a large number of parasitic worms in their bodies. They are not always really dangerous to health, except in children. But it is always advisable to get rid of them as soon as possible; and a doctor should always be consulted.

The *ascaris* is the worm which is most commonly found in children. It is much like an earthworm in shape, and is sometimes mistaken for one. But on looking at it, we observe that if it has the same shape, it can neither move so rapidly, nor lengthen and contract in the same manner as the earthworm. The ascarides live in the digestive tube of animals as well as man. Children contract them more frequently than grown up persons, from their habit of putting in their mouth anything they find, and eating fruit which they pick up on the ground. They often swallow the eggs of ascarides at the same time, which hatch and grow in their stomach.

The *tœnia*.—This is also called the *tapeworm*, and it is erroneously supposed that only one can exist in the same person, although several may occur at once, as is indeed often the case. This worm is flat like a ribbon, and is composed of a series of small square pieces attached to each other, which are called *rings* or *segments*, in spite of their flat shape. The worm's head is at most not larger than a pin's head; it is rounded and exhibits four hollows on the side which are arranged regularly, and are called *suckers*. The first segments behind the head are still smaller than this; but they become larger and larger, and the last are nearly half an inch broad by as much long. There are sometimes a large number of these rings; there may be several hundreds of them, and the entire worm generally measures several yards. All the rings grow behind the head one after the other, so that the last is always the oldest. The head continues to produce fresh ones as long as it lives. For this reason when a piece of the tapeworm has been voided, it is necessary to look if the head is there; for if the head has remained in the body, it will form new segments, and at the end of some time will reproduce a worm as long as the piece discharged.

Isolated segments of a tænia detached from the extremity of the animal are sometimes discharged. The presence of one or more tænias in the intestines may affect the health; but it is a mistake to suppose that it eats part of the food which is taken. The tænia has not even a mouth with which it could swallow anything.

The *flukes.*—Flat parasitic worms which crawl like slugs are often found in the liver of sheep; some are very small, and only a quarter of an inch long; and others measure an inch or an inch and a half in length and resemble a leaf; they are called *flukes*. We also meet with small black or grey worms, very similar to flukes, in marshes, which crawl rapidly on aquatic plants or on the surface of the vessel in which they are placed. These animals are called planariæ, and are remarkable because if they are longitudinally divided in two, as far as the middle of the body, with a sharp pair of scissors or a razor, each half completes itself, and the animal has soon two heads or tails, according as it has been divided in front or behind; and if we then finish dividing it, we have two complete and separate animals instead of one.

*Worm of the staggers.*—Parasitic worms do not live only in the intestines. We have just seen that the flukes are found in the liver; others are found in the brain. Sometimes sheep are attacked by a disease well known to shepherds, and which causes them to turn round instead of walking straight. This complaint is caused by a worm of very peculiar form which lives in the brain of the sheep, and which is always found in animals which have died of this disease. It is shaped like a bladder full of water as large as a nut or a walnut. It does not move, and is not at first recognised as a worm, but nevertheless is one. When this bladder is opened, several white prolongations are noticed in the interior ; and if one is torn with the point of a pin, we find a head just like that of the tapeworm in the centre of each.

Other worms live in the flesh and often in the lard of the hog. They also are bladder-shaped, but not larger than peas or at most than nuts. Like the worm which produces the staggers, these also have a concealed head exactly like that of the tænia. But the most formidable parasite of the hog is the *trichina spiralis,* which is found in the muscles, and when taken alive into the human body in any number, has frequently caused speedy death. All worms which thus live in the flesh of animals may become dangerous to man when swallowed, if the meat which contains them has not been sufficiently cooked. All meat used for food ought to be well cooked ; and badly cooked, or only smoked meat should be avoided. All meat which is suspected to be diseased, or to contain worms, ought to be well boiled for a considerable time, after which it may be eaten with impunity.

# SUB-KINGDOM MOLLUSCA.

This sub-kingdom has been divided into several groups, of which the most important are the *Cephalopoda*, which have tentacles which they use for feet (cuttle-fish) ; the *Gasteropoda*, which crawl on their belly (slugs, snails) ; the *Acephala*, without heads, enclosed in bivalve shells (oyster, mussel).

The animals of this class have neither vertebra, like the vertebrata, nor segments like the articulata. Their skin is smooth, as in slugs ; but they are sometimes enclosed like snails and oysters in a shell varying in hardness. Molluscs generally breathe by means of branchiæ ; but some among them, as the slug for instance, breathe by a lung which opens on the side, behind the head.

The *slug* is a mollusc which will give the best idea of the animals of its class. It crawls about in damp weather or in the evening, and moves by dragging itself on what is called its *foot*. Its whole body is covered with a viscous mucous which leaves a trail behind it. It devours plants and fruits, and does great injury in the garden.

The *snail* is very different from the slug. It walks like it by creeping on its foot ; but it has a shell into which it retires, when danger threatens, or to pass the winter. The snail's shell thus serves it for a house ; but it forms part of itself ; it is attached to its skin, and it cannot go out of it, as is sometimes supposed.

Snails also eat fruits and plants, but they are themselves eaten in some countries, and are said to be very beneficial to consumptive patients. Those which have fed on vine-leaves are said to be the best eating.

The *limnea* or *water snail* has a convoluted shell like the snail, and much resembles it ; it lives in ponds, and although an aquatic animal, it breathes like the slug and snail by means of a lung which it comes to the surface of the water to open. A limnea can easily be seen breathing by merely placing it in a glass of water. In spring the limneæ lay masses of eggs on aquatic plants or even on the surface of the vessels in which they are reared. These masses are about an inch long and a quarter of an inch broad. They are formed of a transparent jelly, and are perfectly transparent ; the nuclei are easily to be seen. These nuclei gradually enlarge, and when the small limneæ in their eggs have not yet attained the size of a pin's head, they may be seen to turn round and round, without this movement ever stopping.

Watersnail.

Many mollusca instead of having a spiral shell, like the snail or the limnea, have one formed of two pieces joined by a hinge, and which are called *valves*. These shells, the oyster, cockle, mussel, etc., are called *bivalves*; and the others are called *univalves*. The animal sometimes opens the valves of its shell, and sometimes closes them, and is thus completely protected by them. When it is dead, the two valves are always half open. To keep them closed, the animal has a muscle which connects the two, and which must be cut to separate them. In opening an oyster, the edge is broken with a knife, but this is not sufficient, and the muscle which holds the valves together must be cut through before the shell can be forced open. The muscle must then be cut under the animal in order to detach it. The points where the muscle is attached to the two valves are indicated by recognisable marks.

M

All the bivalve molluscs do not live attached to rocks like the oyster. The cockle and many others move from one place to another. So does the fresh-water mussel, which is common in streams and ponds. The oysters, on the contrary, always live fixed in the place where they have attached themselves after their birth. The true sea-mussels anchor themselves by means of filaments which they can reproduce when they have been torn. These filaments are called collectively the *byssus.*

*Pearls.*—When eating oysters, we occasionally find small pearls in them, which are sometimes quite round, and sometimes of a less regular shape; they are of no value; but true pearls are also found in a large kind of oyster called the *pearl-oyster*, which is only found in tropical seas, especially on the coast of Ceylon. These pearl-oysters are always found at great depths, and are only procured by very skilful divers. In order to sink quicker to the bottom, they put their feet in a loop fastened to a stone which pulls them down; they remove the shells which they find with a knife, and put them in a basket fastened to a cord : and then rise to the surface to breathe. True pearls are always very hard, while imitation pearls made of glass are crushed by the slightest force.

The shell which produces pearls is itself employed in the arts under the name of *mother-of-pearl.*

*Cuttlefish.*—We sometimes pick up a flat oval white object on the sea-shore, which is hard on one side, and friable on the other,

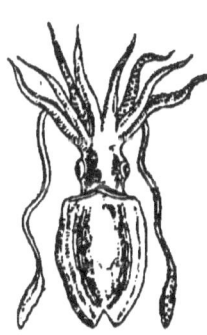

and which is very light. This is called the *bone* of the cuttle-fish; and is used to make tooth-powder. The bone of some species is long, and is called the *sea-pen.* The cuttle-fish are molluscs which have no shell except this, which is found under the skin of their back. . The cuttlefish have arms round their head, provided with a number of small suckers which adhere to any object which they wish

Cuttlefish.  to seize. The mouth is placed in the middle

of these arms, and is furnished with a horny beak remarkably similar to that of a parrot. When these animals are alarmed, they discharge into the water a black fluid resembling ink, which is employed in the manufacture of the so-called *Indian ink*, and of the pigment called *Sepia* by artists.

Although our native cuttlefish have no external shell, yet some foreign species known as the *Paper Nautilus* and *Pearly Nautilus* are celebrated for the beauty of their shells.

# SUB-KINGDOM RADIATA.

This sub-kingdom includes animals of very simple organisation, all the parts of which branch from a common centre.

The most remarkable of these animals live in the sea. They are generally formed of hard or horny parts, and very soft parts which cover these, such as the *coral, sponge, &c.*

Red coral grows in branches attached to the rocks. They fish for it in the Mediterranean with nets of iron wire, which catch it in their meshes and break it. When the coral is alive, its branches are seen to be covered with pretty star-like flowers with eight petals. But on looking at these supposed flowers, we soon see that they move, and that their petals open and close, and lengthen and contract. Each of these is an animal, and it is this which gradually forms the coral, as the oyster forms its shell. Coral therefore only resembles plants because it has branches, but it is in reality an animal. The kind used in trade is of a fine red colour. But there are many other corals in the sea which are white, and are called *brain-coral, sea-mushrooms, &c.*, from their fancied resemblances to various objects.

*Sponge.*—Another group of these singular creatures is the sponge, which is found at the bottom of the sea, and is also an animal. There are many small species of sponge found in the British seas; but the kind used for household purposes is found, like the red coral we have just mentioned, in the Mediterranean. When it is found in the sand, it is heavy, and full of a glairy flesh which is allowed to rot, and is the living part of the sponge. When all this has been removed by repeated washings, the horny, solid framework remains, which supported the flesh of the animal ; and it is this horny part alone which is used.

The *sea-anemones* are closely allied to the corals, but have no hard skeleton ; they live fixed to rocks, and when they open, and unfold their beautifully coloured tentacles, they appear like moving flowers.

The *medusæ* or *jelly-fish* somewhat resemble the anemones, but live floating on the sea. Some sting severely ; but they consist almost entirely of sea-water ; and even the largest species, though measuring two or three feet across, will dissolve so completely as to leave no more than a film on paper.

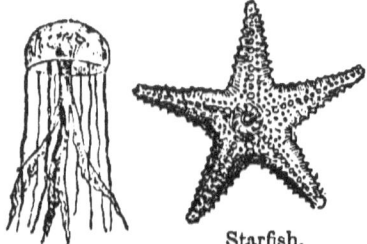

The starfish, which have five or more arms radiating from the centre of the body, and the *sea-hedgehogs*, or *sea-eggs*, which are round and covered with spines, are also radiata.

Some authors also place in this sub-kingdom those swarms of exceedingly

Starfish.

Medusa.

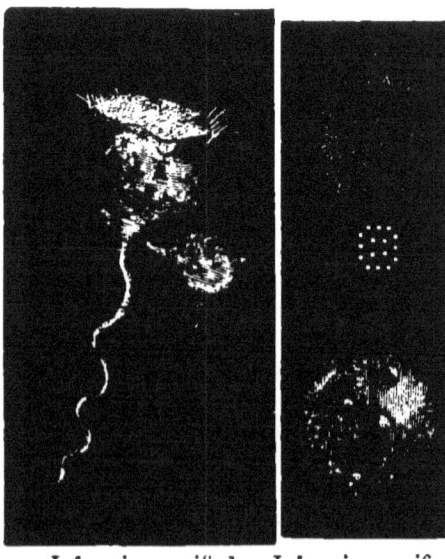
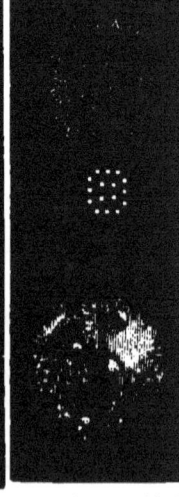

Infusoria magnified
150 times.

Infusoria magnified
300 times.

Infusoria magnified
100 times.

minute animals called *infusoria,* or *animalcules*; but which can
only be seen by the assistance of a powerful microscope; they
principally live in stagnant water.

# VEGETABLE  KINGDOM

———

THE vegetable kingdom includes all plants. These are as different among themselves as animals ; to see this, we have only to compare a patch of mould on a pot of preserves, with an oak or a fir-tree ; and it is therefore necessary to divide vegetables like animals into large divisions and classes.  But before speaking of these, we ought to consider the various parts of which a plant is composed, by taking examples from among the commonest plants.

In order to study what a plant is, we may set some beans in a flower-pot to see them grow.  While waiting for them to appear above ground, let us examine how a bean is formed, and for this purpose we may put one to swell and soften in water for a day or two.  If we then open one, we first find a *skin* or *envelope* entirely separate from its contents.  The interior itself is formed of two halves which are very easily separated, and which are only united by a point.  These two halves are called *cotyledons.*  If the bean is sufficiently softened, we shall see, when these are carefully opened, a very small plant to which they adhere, pressed between them ; a kind of bud is already visible at one extremity, and a root at the other.  Between the bud and root, the little plant is united to the cotyledons in such a manner as to form part of them.  If we now follow day by day the growth of our beans in the flower-pots, this is what we shall observe ; the cotyledons swelling in the damp earth, finally split the envelope; the small plant then frees itself, though still remaining attached to the cotyledons; the bud rises into the air, and becomes the *stem* and leaves; and the root strikes into the earth. The two cotyledons still remain attached to the plant between the stem and the root for a short time like two large leaves, by a thick stalk, but afterwards fall off.

Plants have never more than two cotyledons, but a great number have only one. Plants which have two are called *dicotyledons*, and those which have only one, *monocotyledons*; while plants which have none at all are designated *acotyledons*. Acotyledonous plants differ very much from others; they are mosses, fungi, mould, &c. Dicotyledonous and monocotyledonous plants also differ from one another, so that in most cases there is no occasion to count the cotyledons of the seeds to know in which class they should be placed, for the trunk, leaf and flower show plainly if the plant belongs to one or other of the two divisions, as we shall see by the examples which we shall give further on.

The *trunk.*—When the stem of the plant is of large dimensions, it is called the *trunk.* Trees have a trunk, which is sometimes so large that a man can get into it when it is hollow.

If we saw through the trunk of a tree about a yard from the ground, we first notice that it is formed of three parts; the bark, the wood, and lastly the pith in the centre. The pith is very light and soft. The hollow which contains it, or the *medullary sheath* is often very small in timber trees, but it is sometimes much larger, as for instance in the branches of the elder, where the pith fills more space than the wood.

The *wood* itself forms two layers which can easily be distinguished; the middle layer is darker than the other, and much harder. This is called the *heart* of the wood. The outer portion is less dark and hard. The difference between the heart of the wood and the outer layer or *sap-wood*, is more conspicuous in some woods than others; *Ebony*, for instance, is the heart of a tree the outer layer of which is white like deal.

On looking afresh at the section of the trunk of a tree, we notice concentric rings around its pith, which increase in number with the age of the tree. It was formerly supposed that one of these layers was formed under the bark every year, and that their number would indicate the age of the tree; but it is now known that two or even three layers may be added in one year when the tree is growing rapidly.

The trunks of all trees are not broad near the ground, tapering off to the last branch. Some like the agave, but more especially the palm-tree, grow to a great height, but are as slender below as above. They only grow in length, and have no concentric rings; and the wood is not hardest in the centre, but under the bark. Trunks of this description are only found among monocotyledonous plants. These layers which grow annually under the bark of trees, gradually cover up notches which have been cut into the wood, or nails which are driven into it. If for instance a nail is driven into a tree which has only twenty concentric layers, as far as the outer part of the pith, it will remain there, because the layers of wood once formed do not alter. But other layers will gradually be added to the first twenty, and afterwards when the tree is cut down, we shall find the nail in the heart of the wood, without its having stirred from its place. We can also find again in the wood designs cut into the outer wood, and which have thus been gradually covered by a layer of fresh wood every season.

The stem or trunk is continued downwards by the root. This has often a conical shape, and strikes into the ground like a pivot; such are the salsify and the carrot. All trees have roots of this kind. When they are torn up, the central root is seen to separate into a number of branching filaments. These are the organs by which the root pumps up the water from the earth; and when a tree is transplanted, they ought to be damaged as little as possible. Dicotyledonous plants alone have a root of this kind; monocotyledons never have; and their root is always composed of fibres which are all of equal size, and which all separate from the stalk without ever ramifying. Such is the root of wheat or leek.

The stem raises the leaves into the air; they are generally renewed every year; the stalk which supports them is called the *petiole*. The surface of the leaf is supported by the nervures; to see them well, an oak-leaf may be dried in a book, and then rubbed gently with a brush. All the parenchyma between the nervures fall off, and only the latter remain. Only dicotyledonous plants have nervures like those of the oak. Monocotyledons have none, and the leaf seems to be formed of parallel fibres only, as may be seen in the leaves of the lily, the leek, the wheat and the reed, which are monocotyledonous plants.

Ordinary leaves are called *simple leaves*, but there are also compound *leaves*; and the separate parts of these are called *leaflets*. The horse-chestnut has four or five leaflets, springing from the end of the same petiole; which together form a leaf. In the acacia and rose tree the petiole

Chestnut leaf.        Rose leaf.

terminates in one leaflet, and other leaflets are arranged two and two on opposite sides; which also together form one leaf. To see if the leaf is simple or compound, we must observe the position of the bud which is at the axis of every petiole. When, as in the acacia or chestnut-leaf, the bud is at the axis of the common petiole, then all the small leaves which it bears are only leaflets. They all fall off together with the common stalk.

# RESPIRATION AND NUTRITION OF PLANTS.

The root, stalk and leaves serve for the respiration and nutri-tion of plants ; for plants like animals, breathe and feed. They breathe air by their leaves ; and they feed on the water which they pump up by their roots, and which always contains a great number of salts and other substances in solution. We have seen that in breathing, animals absorb the oxygen of the air, and breathe out carbonic acid. But the breathing of plants is just the reverse ; plants absorb the carbonic acid of the air, and reject the oxygen. Therefore plants purify the air we breathe and it is partly for this reason that it is more healthy to live in the country than in town.

Nevertheless, cut plants in a room may become disagreeable, on account of their odour, which makes some persons ill ; and it is injurious to sleep in a room which contains bunches of flowers.

The water absorbed by the roots is changed into sap, and this rises from the roots to the leaves. It is enough to cut a branch of some trees in spring to see the sap flow in abundance. The sap is often sweet.

But sap is not always the only liquid which flows in vegetable tissues. When we break the stalk of a spurge, of a poppy, or a dandelion, a milk-white or yellow fluid exudes. This liquid has a very acrid taste, and is very different from the sap ; it is called the *milk* of the plant. Some of these are used in pharmacy and in the arts ; opium, india-rubber, and gutta-percha are the milky juices of the poppy, and of some trees which grow in warm countries. This milk, though generally more or less acrid and poisonous, is not always so ; and the milk of some tropical trees is sweet and nourishing.

The *flower* is that part of the plant destined to produce the fruit and seed from which a fresh plant will spring. Many flowers are beautifully coloured, but this is not always the case ; they are sometimes very small and inconspicuous. The nettle, hazel, and oak have flowers ; but they are inconspicuous, and not adorned with bright colours.

In order to study the different parts of a complete flower, we shall take the rocket for an example. We will notice its peculiarities ; and then in studying the different families of plants, will observe what differences they present. If we have not a rocket in flower, we can just as well study the flower of the rape, the turnip, or the wild mustard ; for all these are plants of the same family, and very similar ; and we only mentioned the rocket in preference because it is a little larger. But we must add one caution ; a garden rocket should not be chosen, because cultivated plants nearly always lose their characters.

Imagine us looking at the flower of a rocket. The first thing we see at the end of the stalk which supports it, are four outer petals which surround the bud before it opens. These form the *calyx* which is sometimes simple, or composed of a single piece surrounding the whole base of the flowers ; and sometimes compound. It is generally coloured something like the leaves.          .

Within the calyx is the *corolla*. This is the brilliant portion of the flower ; which is composed of four petals in our rockets ; the wild rose has five. There may be a larger number, or else only one, as in the blue-bell. The corolla is also said to be simple or compound. The parts which form a corolla are called *petals*.

If the calyx and corolla are removed, the essential parts of the flower remain ; the pistil in the middle, and the stamens around it.

There are six stamens in the rocket. They are formed of a filament or thread which supports a kind of yellow bag. When the flower has come to maturity, this opens, and discharges a powder which adheres to the fingers. In the lily, for example, it is very abundant. This dust is called the *pollen*.

In the midst of the stamens is the pistil. It is simple, swollen at the base ; and this which is the most important part, is called the *ovary*. When the ovary is opened, we can already perceive that it is this which will become the fruit when the flower has faded. The fruit is in reality only the ripened ovary. When the plant is still in flower, we can perceive some small white points in the ovary, which are the future seeds. Above the ovary is a stalk with a swelling at the top, on which the pollen falls. Otherwise the ovary does not ripen, and there is no fruit, and consequently no seeds.

But all flowers are not so well formed as that of the rocket. Many have no calyx, as the lily and the tulip; others have no corolla, but only a calyx. The stamens are more or less numerous, and there are sometimes several pistils. Again, the pistils and stamens do not always exist in the same flower, nor on the same plant. The plants and flowers which bear stamens are then called *male*; and those which bear the pistils, and are consequently the only ones which bear fruit, are called *female*. The hemp is one of these plants with separate sexes; and the stalks which bear the female flowers and which produce seed are larger than those with male flowers and stamens.

*Fruits.*—When the ovary has arrived at maturity, it becomes a *fruit*, as we have said. There are a great many varieties of fruits; some enclose several seeds, and others only one; some are fleshy or pulpy, and others hard; and sometimes the fruit seems to consist of the seed only, as in wheat, barley and oats.

In the apple, the pippin is the seed, and there are several in a single fruit. In the date, the seed is the stone, where there is only a single seed to each fruit. It is the same in the peach and plum; but here the seed is only the kernel of the stone; the shell of the kernel forms part of the fruit.

In the poppy, the fruit is hard and horny; and is then called the *capsule*. The seeds are lodged in this capsule, which may have several valves or cavities. Beans and peas are seeds contained in a capsule with two valves and a single cavity. On the contrary, the fruit of the rocket and the rape, is a capsule with two valves and two cavities separated by a partition. The fruits of the gooseberry and currant are called *berries*; and the seeds are placed in the midst of a watery pulp. In the strawberry, the seeds, like those of wheat, are arranged round a fleshy mass, which is eaten, and is therefore called the fruit.

*Sleep of plants.*—The life of plants is as interesting to study as that of animals, and on closely observing them, we notice in them a number of curious actions. Thus many plants sleep at night. In the evening the leaflets of the acacia gradually contract two and two together, and the common stalk droops a little. Next morning the stalk rises, and the leaflets unfold again for the whole day. The sleep of clover is equally remarkable. On the approach of evening, the two lateral leaflets of the three which compose its compound leaf fold together, while the middle one folds over the two others, and covers them like a roof. In the morning, the three leaflets unfold afresh.

# CLASS OF DICOTYLEDONS

## OR PLANTS OF WHICH THE SEED HAS TWO COTYLEDONS.

## FAMILY UMBELLIFERÆ

The family of umbelliferæ includes a great number of useful or aromatic plants, such as the *parsnip*, the *anise*, the *caraway*, the *chervil*, the *parsley*, *celery* and *carrot*; and other plants which are ·poisonous, such as the *hemlock*, and the *hemlock water dropwort*. The umbelliferæ may easily be known, because all their flowers are arranged so as to form a kind of parasol or *umbrella*, whence they derive their name; the stalk terminates suddenly, and radiates from this point a number of small stalks often surrounded at the base with a sort of frill of leaves. All these short stalks divide in their turn a little higher up, like the principal stalk; and there is also a second frill of leaves. Each of these divisions bears a flower. This has a corolla with five petals and five stamens. The corolla and stamens are inserted into the ovary itself, as is the case in many plants. The flower of the umbelliferæ is always small.

The stalks of the umbelliferæ are often hollow; and there are no trees in this family. The flowers of the elder seem also to form an umbel; but it is easy to see that they have not the same characters.

The *carrot* grows in Britain in a wild state, but the root is not so large, nor red as it becomes when cultivated. The leaves of the carrot are compound, with a great many divisions. The stalk of the carrot, like that of other umbelliferæ, dies every year when the seeds are ripe ; but the root still lives, and throws up a new stalk in the following year. Each flower, as in all umbelliferæ, bears two seeds ; those of the anise and caraway are very aromatic. On the contrary, in the parsley, chervil, and angelica, it is the leaves and stalks which are valued for their flavour.

Poisonous umbelliferæ may generally be known by the dis-agreeable smell which is perceived on rubbing one of their leaves in the hands. The hemlock is a large plant with reddish spots on the stem. The fools', parsley is much commoner, and care must be taken not to mistake it for the real parsley. It may always be known by the shape of the leaf frill at the base of the last stalks which support the flowers. This frill is composed of three leaves as narrow as threads, which are drooping, and all three placed together on the same side of the stalk.

# FAMILY SOLANACEÆ.

The family of the Solanaceæ, like that of the umbelliferæ includes at once plants which are among the most useful to man, as the *egg-plant*, the *tobacco*, the *tomata*, the *potato* and the *capsicum* ; as well as the most poisonous plants, such as the *night-shades* and the *thorn-apple*. It is true that some of these are of great service in the hands of physicians, and then become useful plants. The Solanaceæ sometimes grow to small trees. Their flowers differ considerably ; from that of the potato, which is wheel-shaped, to that of tobacco, which is bell-shaped. But the calyx is always simple, with five indentations ; and the corolla is similar. Lastly there are five stamens, which in the potato and the night-shades are perforated with a small hole, through which the pollen escapes.

The potatoes which we eat are swellings which are formed at different points of the long creeping roots of the plant. These swellings are called *tubers* ; they are formed by the large quantity of starch which collects there. The starch in potatoes may be extracted by crushing them, and afterwards sifting the bruised pulp in water. The grains of starch being very small, pass through the seive, and sink to the bottom of the vessel by their own weight.

If a potato-plant is left without being pulled up, the plant dies at the end of the year, but next year each potato produces a new plant from the hollows which it contains, and which are called *eyes*. A field of potatoes is always planted with pieces of potato which still contain the eyes. If potatoes are kept in a damp and dark place during the winter, these eyes are seen to bud, and produce long pale filaments, on which pale yellow leaves some-times grow. They are of a yellow colour because they have grown in darkness ; and it is always necessary for the leaves of plants to acquire their fine green colour, for them to be exposed to the sun, or at least to daylight. The hearts of cabbages and lettuces are not so green as the leaves surrounding them ; and in the same way, the earth is heaped round celery, that the leaves may remain yellow, and that the light may not turn them green. The parts which thus remain yellow are also more tender.

In the *tobacco*, the leaves are used for the very general practice of smoking, chewing, and snuffing. The leaves are pulled off, and allowed to dry ; when they are yellow, they undergo some preparation, and are then rolled up to make cigars, or chopped to make smoking tobacco, or powdered to make snuff. The use of tobacco is never indispensable ; it can always be given up, and it is often injurious to those who are addicted to it. The flowers of the tobacco are arranged in pretty clusters, like those of the

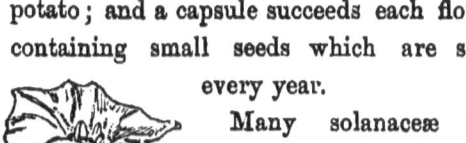

Tobacco.          Tobacco flower.

potato; and a capsule succeeds each flower, containing small seeds which are sown every year.

Many solanaceæ bear berries which are always dangerous to eat, and some are very poisonous. The berries of the *bitter-sweet*, or *woody nightshade*, which grows in hedges, are first green and then red; they have a sweetish taste, which soon changes to bitter. The berries of the deadly nightshade are of a reddish brown, and the calyx of the flower remains adherent to them. The plant, which is one of the most poisonous found in this country, should be very carefully avoided. It is generally very dangerous to eat unknown berries, however tempting they may look. It is easy to perceive when children have been made ill by eating the berries of the deadly nightshade, for the pupil of the eye is very much dilated, so that the iris is scarcely visible. The child should be made to vomit immediately, and a doctor should

Deadly nightshade.

be sent for.

Other solanaceæ which are cultivated have highly esteemed and wholesome fruit, such as the *egg-plant* and the *tomato*, as well as the *capsicum*, the fruit of which is used in pickles.

# FAMILY EUPHORBIACEÆ.

This includes a number of plants which have a very peculiar appearance. The various kinds of spurge belong to this family.

The flower exhibits no striking colours, and there is a large pistil in the centre, from which we can already perceive the shape which the fruit will take. Many Euphorbiaceæ have a milk-white and very acrid juice.

The *box* belongs to this family. It is an evergreen, and grows very slowly. However, it grows to a large tree in time, and its wood is very expensive.

Spurge.

It is used for wood engravings. A picture is drawn on a block of box-wood, which is then cut in such a manner as to leave only the marks of the drawing in relief ; and it is then covered with ink, and printed from.

The *castor-oil plant* also belongs to the family Euphorbiaceæ. It is sometimes grown in England for ornament ; but in hot countries the seed yields an oil which is used in medicine as a purgative.

---

# FAMILY CHENOPODIACEÆ.

This family contains the *spinach*, the leaves of which are eaten ; and the *beet*, which is an extremely useful plant. The plants of this family, like those of the last, bear small and inconspicuous green flowers.

The beet-root is not only used for human food when cooked, and for food for cattle when raw ; it is largely cultivated on the continent for the manufacture of sugar, which is contained in the

N 2

sap of which the beet-root is full. The root is chopped fine, and pressed, and the juice is then allowed to evaporate. The sugar remains, and is then made into loaves, to be sold.

---

# FAMILY POLYGONACEÆ.

The *sorrel*, the *dock*, the *snake-weed*, the *rhubarb*, and the *buck-wheat* belong to the family *Polygonaceæ*. ,The fruit is dry like a grain of corn, but it has three sharp edges, and three sides. The flowers of the Polygonaceæ are not generally more brilliant than those of the two preceding families. The flower of the buckwheat however is an exception; it is white, and has six divisions and six stamens. It is easy to perceive at the first glance that the buckwheat has no resemblance to common wheat; the first is a dicotyledonous plant, and the second has leaves without transverse veins, and belongs to the monocotyledons. Buckwheat is only grown in England in small quantities as food for deer and game; but

Fruit and flower of Buckwheat.

its seed yields a brown flour which is used in some countries for making bread and pancakes, &c.

The *rhubarb* which is used as a medicine is the root of a plant of this family which grows in Asia. The large-leaved rhubarb which is grown in vegetable gardens, and the stalks of which are used for pies, puddings, &c., is an allied, but perfectly distinct plant.

# FAMILY PAPAVERACEÆ, DIAGRAM 10.

The name of this family is derived from the Latin word for poppy. Contrary to the families which we have just been noticing, it includes plants with a coloured corolla, which is bright red in the corn poppy. The calyx is formed of two parts, which fall off when the bud opens. The corolla has four petals, and the stamens are very numerous. The pistil is already very large, and

Poppy flower.

is of the same form as the fruit which will succeed it, the so-called *poppy-head*. Several holes open at the top of the ripe capsule, to allow the seeds, which are very small, to escape. On closely observing them, their surface is seen to be as it were chagrined; they are bean-shaped, and very numerous.

Some kinds of poppy are cultivated for their seed, from which oil is extracted; but the chief use of the plant is in medicine. The poppy-heads contain a substance which produces sleep and relieves pain. In hot countries incisions are made in the stalk of poppies in flower, and a white juice like that of the spurge exudes from them. It turns brown on exposure to the air; it is scraped off and sold under the name of *opium*. This substance is much used by doctors : and it has the property, like the poppy-head, of causing sleep, and relieving attacks of pain.

The *celandine*, which grows by road-sides, belongs to the poppy family, and may be known by its yellow and very acrid milk, which is sometimes used to burn off warts.

Other curious plants of the same family are the white and yellow water-lilies which unfold their leaves and flowers on the surface of ponds and slow rivers. The buds grow at the bottom of the water, and rise to the surface to open. But the flower closes every evening, and sinks under water for the night. Next morning it rises again to the surface, and opens afresh till evening.

# FAMILY RANUNCULACEÆ.

The family Ranunculaceæ includes plants which have all brilliant corollas composed of several parts ; but which are very different in external appearance.  The *hellebore*, the *Christmas-rose*, the *monkshood*, the *larkspur*, the *clematis*, which grows in hedges, and the *buttercup* or *ranunculus* are only alike in having their fruits formed by a cluster of several small dry seeds.  Moreover, nearly all the ranunculaceæ are poisonous plants, which cattle often avoid, and which make them ill if they eat them. But when dried they seem to lose part of their various properties ; and it is for this reason that the buttercup does not injure the quality of hay.  The calyx of the buttercup is yellow ; its corolla has five petals, and a great number of stamens like the poppy ; but instead of having only one pistil, it has several, each of which becomes a dry seed when ripe.

---

# FAMILY LEGUMINOSÆ, DIAGRAMS 10 & 12.

The family leguminosæ is one of the most important in the whole vegetable kingdom.  It includes a great number of plants, from herbs to the largest trees.  Its name is derived from its fruit, which is called a *legume* in botany.  This fruit is a capsule with two valves, and is called a *pod* ; and the seeds are inserted upon one of the lids.  All the leguminosæ bear a similar fruit. This large family includes the *furze* and *broom*, which are not very useful ; and other plants which are used for the food of

Pea.

man ; such as *peas*, and beans of all kinds ; lupins and lentils.

Other plants are used for forage, such as the *lucerne*, the *vetch*, and the *clover* ; one plant yields oil, *arachis* ; others are large trees, such as the *laburnum*, the *liquorice*, the *indigo*, the *mahogany*, and the *carob-tree* ; and the curious *sensitive plant* also belongs to this family.

All the plants of the family leguminosæ have a pod for their fruit, and their flower is also very similar ; the calyx is very regular, with five divisions ; the corolla is irregular, and formed of five parts ; it is often beautifully coloured, and is easily recognised by its shape, which has been compared to that of a butterfly. There is an uneven part erected above, and then two lateral parts which represent the wings. The stamens are numerous, but are divided into two distinct groups ; one is placed by itself ; others are soldered together, and surround the pistil, which is already pod-shaped. The leaves of leguminosæ are almost always compound.

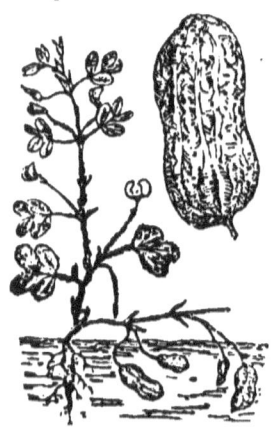

Arachis.

The *arachis* or *earth-nut* is very much like clover, but will only grow in hot countries. When the flower is over, and the fruit begins to ripen, the stalk which bears it bends down, and the fruit buries itself in the ground where it ripens. It is gathered there, and brought to Europe in large quantities, and is crushed to extract *ground-nut oil.*

Liquorice juice is extracted from the root of a shrub which is very common in South Europe. It is dried, and formed into sticks ; and the root itself is also sold, and is chewed for its pleasant taste.

The *indigo-plant* yields one of the most useful dyes, and only grows in the hottest countries. To extract the indigo, all the green parts are put in cellars, where they decay rapidly. The indigo then separates, and is collected in the form of a blue paste which is made into small lumps like pieces of chalk, which are exported into all parts of the world. The indigo is one of the best dyes for its fastness, and one of those which are most easily renewed.

The *sensitive plant* is a small leguminous plant which only grows in hot countries, but is often grown in greenhouses because of the curious sensibility to which it owes its name. In the evening, its compound leaves close up and hang down to sleep ; but those of nearly all leguminosæ act in the same manner. We have already spoken of the sleep of the clover, and of the false acacia. What is peculiar in the sensitive plant is that when the leaves are expanded and the weather is warm, the leaves close and droop immediately if touched, as if they were going to sleep. Little by little they rise up and re-open slowly, to close and droop again, as soon as they are touched anew.

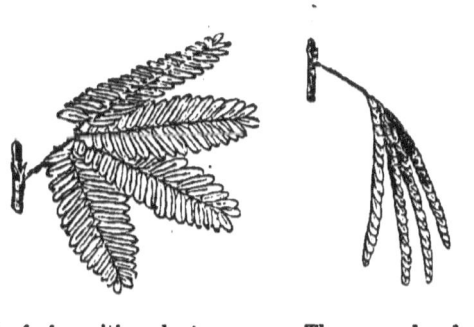

Leaf of sensitive plant open.　　The same closed.

# FAMILY LABIATÆ.—DIAGRAM 12.

The Labiatæ like the Leguminosæ form a family in which there is a very considerable resemblance between the plants. We may take the *white dead nettle* as an example, though it is not a useful plant to man. We notice immediately that the stalk is square, unlike that of most plants ; moreover the leaves are always arranged two and two on the stem ; they are *opposed*, as it is said in similar cases. The calyx is formed of a single piece in five divisions; the corolla is also formed of a single piece, but it is very irregular; it has a peculiar form, and seems to have two lips, an upper and lower one; and there are four stamens, which are remarkable for two being large and two smaller.

Nearly all the plants of the family labiatæ are fragrant, and none are poisonous. The principal plants are the *rosemary*, the *sage*, the *mint*, the *lavender*, the *thyme*, the *marjoram*, and the *balm-mint*.

# FAMILY RUBIACEÆ.—DIAGRAM 12.

This family is one of the most useful to man. It includes three plants of the greatest importance, the *cinchona*, the *coffee*, and the *madder*. *Quinine* is the bark of the cinchona, a large tree which grows in South America; the coffee also is a tropical shrub largely grown in Southern and Western Asia; the madder is cultivated in South Europe; the only plants of the family Rubiaceæ which are found everywhere are the *yellow bed-straw* and the *goose-grass*. The last is much like the madder. Its stalk is square like that of the Labiatæ, and the leaves also grow on the same level, but more than two together. The corolla is regular, with four or five divisions and as many stamens. The fruit is a double pod.

Quinine or Peruvian bark is of two kinds, the grey and the red. It has the property of curing intermittent fevers, which are especially frequent in marshy countries. Instead of using the bark itself, a salt called quinine, which is extracted from it and possesses the same qualities, is employed. Indeed, the bark only owes its virtue to the quinine which it contains.

Coffee is the seed of a shrub which grows in Arabia, and is cultivated in all warm colonies. There are two seeds, or *coffee-*

*beans* in each pod which the flower produces. The beans have no flavour; and this is only developed when they are roasted. Coffee is one of the most wholesome beverages, and does not intoxicate like beer or wine. It ought not however to be taken in excess because it then produces trembling, and unfits us to perform various duties with sufficient skill.

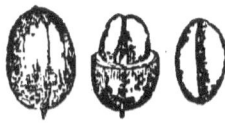

Coffee bush.                    Coffee bean.

The madder is cultivated in South Europe, and yields a fine dye ; and it is this which is used in the manufacture of military clothing. The root, which resembles a small branch of hard wood, is the part employed. The bark and pith of the root are of a dark red, and the wood is yellowish.

The name of the family Rubiaceæ is derived from the Latin name of the madder.

---

# FAMILY URTICACEÆ.—DIAGRAM 15.

The family Urticaceæ is another which has not brilliant flowers, but which is one of the most useful to man. Although on the one hand it includes the common nettle, which stings severely, and the pellitory which is of no use, it also includes the *fig*, the *mulberry*, the *hop*, and the *hemp*. We thus see that some of the plants of this family are herbs, and others large trees. There is nothing remarkable about the flowers, which have generally separate sexes. These sometimes grow together on the same plant, and sometimes grow on different plants, as is the case, as we have mentioned before, with the hemp. The fruit of the Urticaceæ varies much.

In the *nettle*, the male and female flowers grow on different plants. The leaves are studded with very stiff hairs, which pierce the skin, and inject a very active venom with which they are filled ; and it is on this account that the sting is so painful. The nettle may however be handled with impunity if suddenly and firmly grasped, as the hairs will then be bent or broken. The stalk of the nettle is fibrous like that of the hemp, and it has sometimes been proposed to use it for manufacturing purposes.

The hemp has the male and female flowers on different plants. We have already said that the male plant is smaller than the female. The male flowers have five stamens, and the females have a pistil which becomes the *hemp-seed* used for feeding caged birds.

The hemp possesses very stimulating properties, and an allied plant, the so-called *Indian hemp*, is used in some countries to produce a pleasurable intoxication. It is sometimes used in medicine.

Hemp is generally sown in spring, and when the time has come to harvest it, the plants are pulled up, tied in bundles, and allowed to decay either in the ground, or in water. This is done to rot away all parts of the plant which cannot be used for making cordage. The hemp then undergoes two other operations ; *stripping* and *combing*. In the first, the central part of the stalk which is not fibrous is broken off, and in the second, all the refuse is removed ; and the tow remains, which is first transformed into thread, and then into canvas and cordage.

The *hop* has leaves divided like the hemp, and is a climbing plant which grows in hedges, and is also largely cultivated in the south of England to be used in making beer. Beer is made of barley, but is then sweetish, and hops are added to it to give it its

Female flower of hop.      Hop.

proper bitter taste. The hop is also a plant with separate sexes. The female flowers grow in clusters at the end of the stalks, and somewhat resemble small pine-cones, with very thin and delicate scales. We shall come to families in which all the flowers are thus arranged in similar clusters. It is the female clusters of the hop which are employed.

The hops are planted at the foot of long poles six or eight yards high, up which the plant climbs. When the time to gather it has come, the year's growth is cut, the pole is lowered, and the clusters are gathered and dried for sale. At the bottom of each scale is found a small quantity of a yellow bitter resinous substance, which gives its taste to beer.

Mulberry.

The *mulberry* is a very valuable tree, not for the sake of its fruit, though this is eatable, but because it is grown so extensively in Southern Europe for feeding silkworms.

The *fig* is also an abundant tree throughout Southern Europe, though only seen occasionally in English gardens. Figs are eaten either fresh, or dried; and dried figs packed in boxes are very largely imported into England.

The fig is a fruit which at once strikes us as peculiar, because the fruit is not seen to succeed the flower on the tree. It is really not a fruit like others. If we pluck a young fig and open it, we perceive a cavity in the interior with an orifice at the top. On

 the sides of this cavity we can distinguish an immense number of very small dull-coloured flowers like those of other urticacæ. These are male and female. The pistils of the latter become the seeds which we find in dried figs. As they ripen, the sides of the pouch thicken, the orifice above closes ; and the well-known sweet fruit is the result.

Fig.

The family Urticaceæ derives its name from the Latin name for the nettle.

# FAMILY LAURACEÆ.

This family is much less important than the preceding, and includes in addition to the laurel from which it derives its name, the trees which yield *cinnamon*, *nutmegs*, and camphor. The plants of this family are therefore, like the labiatæ, chiefly aromatic plants, or rather trees.

The true *laurel* is the tree which is generally called in England the *bay*-tree. Its leaves are used as a flavouring on the Continent. It is an evergreen shrub which grows to a height of thirty feet. The flowers are small and inconspicuous, with four or five divisions, and nine stamens. Their pollen-sacs have small valves, which open to allow the pollen to escape.

In ancient times, victors at the public games, etc., were crowned with laurel. Our term *poet-laureate* owes its origin to this custom. The fruit of this tree is a small berry.

The *cherry-laurel*, generally called the *laurel* in England, is a handsome shrub with large dark green leaves. It seldom grows to a great height. It is an evergreen, and bears a spike of pretty white flowers, but does not flower till it is several years old. Its leaves, when bruised, have a very pleasant odour, which is due to the large quantity of prussic acid which they contain, which renders the cordial called cherry laurel water, which is distilled from them, a most dangerous specific.

*Cinnamon* is the bark of a kind of laurel which grows in the East Indies. The young branches are cut down, and the bark is peeled off and dried, after which it is rolled into the form in which it is known in commerce.

*Camphor* is produced by a large kind of laurel which is found in China and Japan. The camphor exists in the sap ; and when an incision is made in the bark, the liquid which exudes dries, leav-

Cinnamon-tree.

ing camphor. But this method of extracting it, which is similar to that used to procure opium, would be too costly. The broken branches and roots of the tree are simply boiled ; and the camphor then separates from them. It is sent to Europe in greyish masses, which are purified to give the camphor the whiteness which it possesses when sold. One curious peculiarity which camphor exhibits is its behaviour when a small piece is put into water. If the water is very pure, and the vessel very clean, the camphor at the surface becomes agitated, and darts about in every direction ; but if the point of a knife which is a little greasy is dipped into the water, the camphor at once ceases its movements, and will not move again on the same water. When the camphor itself has touched a slightly greasy substance, these motions do not show themselves ; and it is therefore necessary for the success of this experiment that both the vessel and the water should be as clean as possible.

The nutmeg is the fruit of another tree of this family, which likewise grows in tropical countries.

# FAMILY MALVACEÆ.—DIAGRAM 15.

This family contains various low plants, such as the *hollyhock*, *mallow, and marsh-mallow*. But it also includes large trees, and among others, one of the largest known, the *baobab*. One of the  most useful plants to man, the *cotton-plant*, also belongs to this family. The malvaceæ have generally beautiful flowers; the calyx is in a single piece, and the corollæ has five petals. The stamens are very numerous, and soldered together round the pistil, which is thus enclosed in a kind of sheath. The fruit is a capsule, contaning a variable number of seeds.

Mallow.

The cotton plant is a shrub which grows in hot countries. It is a textile plant ; but the material which it yields is not obtained from the stalk as in the hemp, but is found in the fruit. The cotton-plant bears a beautiful yellow flower, and the capsule which succeeds it opens in several valves when ripe. The cotton is then seen inside the capsule, forming a lining packed round the seeds. It is roughly cleared from the husk, and packed in *bales* to be exported to countries where it does not grow. It is an article of great commercial importance. It is first cleaned and carded, to free it from the seeds which still remain in it. It is then spun and woven to make *calico*. When coloured designs are printed on it, it is called *print*. Whole countries live by the cotton trades : both those where it is cultivated, and those where it is spun, woven, or printed. The cotton trade is one of the most important in the world.

Another plant closely allied to the family Malvaceæ is that which produces chocolate. The cacao is a tree which grows to the height of thirty or forty feet, and bears red flowers. It only grows in tropical America· Its fruit is about four inches long, and resembles a cucumber, but the outside is as hard as wood. Each contains twenty or thirty seeds as large as almonds. *Chocolate* is only a mixture of sugar, spices, and cacao beans roasted and ground. These, like coffee beans, only acquire their flavour when roasted. Cocoa is also prepared from the cacao beans.

Cacao-pod.

## FAMILY LINEACEÆ.--DIAGRAM 15.

This small family only contains one plant, but one of the greatest importance, the *flax*. It is very pretty, and would be grown for ornament if it were not also a valuable textile plant. The corolla is of a beautiful cærulean blue, and has five petals ; there are ten stamens, only five of which are provided with pollen sacs. Flax is converted into linen in exactly the same manner as hemp. It is pulled up, steeped, broken and combed, to get rid of what cannot be spun. The residue forms a very fine tow, much finer and more compact than that of hemp, and from which muslins and linens of the finest quality are manufactured.

## FAMILY OLEACEÆ.—DIAGRAM 14.

The family Oleaceæ also contains only one plant of real import-ance, the *olive*. Its flowers are not remarkable. Its foliage is dark, and composed of small, stiff, scattered leaves. The fruit is fleshy

with a stone.   The olive grows slowly, like the box, and its wood
is also very hard.   It is not grown in England, and grows to a
larger size in the East than in South Europe.   The fruit is only
gathered when the cold weather sets in.   The olives are the
pickled to be eaten, or pressed to extract the oil, which is the finest
kind known.

The *ash* is placed in the same family as the olive.   The *manna*
which is sold by druggists is produced by a species of ash which
grows in the East.   The manna flows from the bark of the tree
and forms what is called *tears*.   It is gathered and sold without
any other preparation.

---

# FAMILY ROSACEÆ.—DIAGRAM 14.

The family Rosaceæ is one of the largest and most important.
Besides the rose, the strawberry, and the raspberry, it includes
a great many fruit-trees ; the *apple, pear, peach, medlar,
cherry, plum, apricot*, and *almond*.   The characters of the Rosaceæ
must not be looked for among our garden roses,—they are plants
which have been completely changed by cultivation.   Cultivation
often produces this effect, and it soon happens that a cultivated
plant no longer resembles the plant from which it was derived.
One of the first effects of cultivation is to multiply to an inordinate
extent the number of petals, and to produce what are called *double
flowers* instead of single flowers.   But cultivation likewise increases
their perfume.

The character of the family Rosaceæ, as it is exhibited by the
wild roses which grow in the hedges, is to have five expanded
petals like the oleander, and numerous stamens.   But these petals

o

Wild Rose.

and stamens are inserted on the calyx, which is formed of a single thick fleshy piece. There are sometimes several pistils and sometimes only one, according to whether the fruit is to produce several seeds, like the rose, the strawberry, and the raspberry ; or only one, like the plum and the almond.

The double rose is not simply a garden plant, but is cultivated extensively in some countries to extract from it the precious oil called *attar of roses*. Whole fields are planted with roses, the petals are collected, and they extract by squeezing them a few drops of this essence, which is always very dear, but which has a very powerful odour.

The *apple* is not only the most valuable of all fruits, but in some parts of England it is made into a drink called *cider*. The apples are peeled and pressed, and the extracted juice is allowed to ferment.

# FAMILY CRUCIFERÆ. DIAGRAM 14.

All the plants of the family Cruciferæ have much general resemblance, but most of them are small herbs. It includes the *mustard*, the *rape*, the *cabbage*, the *turnip*, the *radish*, and the *rocket*. We have already described the flower of the rocket in detail (see p. 172, diagram 9); all the others resemble it, and are composed of four petals arranged in the form of a cross ; hence the family derives its name Cruciferæ, " Cross Bearers." The fruit is a capsule with two valves, but has two divisions separated by a partition, while in the fruit of the family Leguminosæ, there is no film.

The Cruciferæ are especially food-plants ; and are also very wholesome. We eat the root of the radish, turnip radish, and turnip; and the large bud which surmounts the cabbage-stalk. The cauliflower is a cruciferous plant with very crowded and aborted flowers, while their stalks have become monstrously developed by cultivation. The seed of the rape is pressed to extract colza oil. The essence which causes mustard to irritate the eyes and tongue, does not exist in the seed itself, and is only formed when water is added to flour of mustard. If we taste flour of mustard before mixing it with warm water, and afterwards, we shall easily perceive the difference.

# FAMILY AMPELIDEÆ.—DIAGRAM 14.

The only plant of this family which we shall mention, is the vine. Its flowers have a calyx with five teeth. The corolla has five petals, but they do not open; they are soldered together at the top, and detach themselves like a small bell, which falls off immediately. There are also five stamens. The well-known fruit is a berry with about four seeds or stones. The vine is universally cultivated throughout Central and Southern Europe; but is no longer grown in England to any extent, except in greenhouses; for the fruit does not ripen well in our cool summers; and scarcely any grape-wine is now made in England, though the vine was largely cultivated for this purpose in the middle ages.

It is sometimes supposed that the difference between red and white wine depends on the colour of the grapes from which it is made; but it really depends on the preparation. The juice is pressed out of the grapes, and allowed to ferment, and thus becomes wine. The refuse is then distilled to make brandy and other alcoholic drinks.

Wine is an excellent drink in moderation, but the use of

brandy and all spirituous liquors is nearly always injurious, except perhaps to revive those who have been exposed to wet or cold, *but even in this case very little should be taken.* This advice is important, for it should never be forgotten that brandy is still more dangerous in winter than in summer.

---

# FAMILY COMPOSITÆ.—DIAGRAM 11.

This is one of the most extensive families among dicotyledonous plants. It does not include any trees, but it covers the fields. Several compositæ are useful, and others are cultivated as ornamental plants. This family includes the *endive*, the *lettuce*, and the *dandelion*, the leaves of which are eatable ; the *salsify* and the *Jerusalem artichoke*, of which the roots are eaten ; the *chicory*, the leaves of which are used as salad, and the powdered root mixed with coffee ; the *centaury*, thistle, and *artichoke*, which are also eaten by man or beast ; the *marigold, dahlia,* and *sun-flower* which grow in our gardens ; the *wormwood*, from which the highly deleterious liqueur called absinthe is distilled ; the *camomile*, which is used in medicine ; and finally the *daisy*.

If we examine all these flowers, we immediately notice that while they have considerable resemblance to each other, they are at the same time very unlike other flowers ; and at first we recognise neither calyx, nor corolla, nor stamens, nor pistil. The fact is that the thistle, sunflower and daisy are not really flowers, but clusters of flowers, as it is easy to perceive with a little attention, and it is on this account that these plants are called compositæ. If we take a dandelion, or a sunflower, and pull out the yellow centre, we shall see that it is composed of a great number of parts, and in each of these a small flower is easily to be distinguished, with its corolla, stamens, and in the middle of them a pistil divided into two recurved branches surmounting the whole. The corolla is inserted on the ovary itself, and this is placed with the other ovaries of the adjacent flowers, very regularly on a kind of plate or receptacle. When we eat an artichoke, we take out what is called the *choke*, which is nothing else than small flowers still in bud, for which the bottom of the artichoke forms the receptacle. This is very large in the sun-flower, and is also conspicuous in the thistle, when its violet flowers are pulled out. The receptacle is always surrounded with scales or leaves like those of the artichoke. They form a kind of basket in which the flowers are contained like a bunch of flowers in a vase.

Sometimes all these small flowers are alike, as in the thistle, but in other cases they differ considerably. Those of the margin are often larger and differently coloured; in the daisy and camomile they are white, while those in the centre of the flower are yellow. They may be regular or irregular, male, female, or hermaphrodite, or have neither stamens nor pistils.

Let us take a thistle first. All its flowers are nearly similar. The corolla is regular, and shaped like a tube widened above, and cleft into five divisions. ˙The corolla is inserted upon the ovary; and the stamens in their turn are inserted on the corolla. They are five in number, and soldered together by their pollen-sacs, while the threads are distinct. The united sacs·form a canal into which the pistil, which ends in two bifurcated branches, passes. The thistle gives us an example of a composite plant, the flowers of which are all regular. When they fade, the ovary becomes a seed furnished with small silky sails which allow it to be carried away by the wind.

The centaury and wormwood have very small baskets, and their flowers are complete and regular like those of the thistle.

In the chicory, the flowers are still complete, that is, they are . male and female, but they are no longer regular; the corolla, instead of being tube-shaped, is cleft from above downwards, and removed beyond the receptacle, in the shape of a small plate, at the extremity of which the five divisions of the regular corolla of thistles are still to be recognised. Each of these corollæ is inserted, as in the thistle, upon the ovary; and it has five stamens with the pistil in the middle. We find in the camomile and the daisy, besides the ordinary yellow regular flowers, others which are irregular like those of the chicory, and white, and form a rim round the others. But these irregular flowers are also incomplete, they have neither stamens, nor pistils, and are neither male nor female, but neuter : and therefore produce no seeds.

In the sunflower the receptacle is also surrounded with irregular neuter flowers; but there is a difference between the flowers within them ; in the centre of the receptacle they have a pistil and no stamens, and are female flowers ; but towards the edges of the receptacle, they are male flowers, with stamens and no pistil. The latter, like the neuter flowers, of course produce no seeds.

The dahlia is a composite plant in which all the flowers have been rendered neuter and irregular by cultivation. In its native country, there is only a row of large neuter flowers round the receptacle, and the others are small and yellow, as in most compositæ.

---

# FAMILY CUPULIFERÆ.—DIAGRAM 16.

This family has inconspicuous flowers, and includes most of our forest trees, such as the *willow*, the *poplar*, the *birch*, the *elm*, the *oak*, the *Spanish chestnut*, the *beech*, the *hazel*, and the *hornbeam*. The flowers are in most cases reduced to a simple scale, sometimes isolated, and sometimes in clusters, like the flowers of the *hop*. The sexes are sometimes found on the same tree, and sometimes separated. They are male clusters, or *tassels*, which hang on the branches of the hazel at the end of winter, before the leaves have appeared. On looking at them closely, it is easy to see the stamens inserted at the base of the scales.

Male Tassel of
the Hazel.

The female flower is often single.   The fruit is variable, but consists in many cases of a seed which seems to be contained in a small cup, or *cupula*, from which the name of the family is derived The cupuliferæ yield valuable timber and bark for many indus- - trial purposes.   The bark of these trees is almost always very bitter.

The acorn of the *oak* is the typical example of a fruit contained in a cup.   It is well known in what esteem the wood of the oak is held by builders and joiners, for its durability and beauty.   The bark is not less useful.   When young oaks are cut down, it is carefully removed, and is used in tanning hides, to convert them into leather. The tan after being used, is sometimes spread in the streets in front of houses where a person is dangerously ill, to deaden the sound of passing vehicles.

*Cork* is the bark of another kind of oak, which grows in North Africa.   When the tree is fully grown, an incision is made at the top and bottom of the trunk, and the bark is removed.   It is allowed to dry, and then cut into corks.

It is on the oak that the *gall-nuts* are produced in some countries by the attacks of the cynips, which are used in the manufacture of ink.   (See p. 144.)

The hazel, or nut-tree has a fruit in a cup like the oak, but this covers it entirely.   The branches of the hazel are straight and flexible, and are used' for many purposes.

The *poplars* and *willows* generally prefer a damp situation. Their wood is light and of little value, but is used for purposes where lightness without great durability is required ; for making cricket-bats, for instance.   Their fruit differs from that of the oak and hazel, and is a capsule which contains seeds furnished with a kind of down : and they are sometimes carried by the wind to a very great distance, when the capsule opens.   The young shoots of the willow are extremely strong and flexible ; they are called *osiers*, and are employed for all kinds of basket-work.

The *birch* may be immediately recognised in woods by the conspicuous whiteness of its trunk, and by its foliage, which is not so thick as in other trees. The birch-bark flakes off out-side, but is very solid, and when a large branch is cut into lengths, very strong boxes can be made of the bark.

The edible, or *Spanish chesnut*, although a native of Asia, is abundant throughout Central and Southern Europe, and in South Europe the fruit forms a most important article of food. It is also not rare in England. Each fruit generally contains two chesnuts. The flowers are inconspicuous, like those of all the family of Cupuliferæ, and are simply composed of small scales. They are remarkable for their very powerful and sickly odour. Chesnut wood is valuable, and it is said that insects will not attack it. The branches are always straight and flexible, and are used, when split, for barrel-hoops.

The *horse-chesnut* does not belong to this family, and is a foreign tree which has been introduced into Europe, like all trees which bear handsome flowers, and are not fruit trees. It is also a native of Asia. It is extensively grown in England for ornament, and is much more frequently seen than the Spanish chesnut. It is a hardy tree of very rapid growth, and presents a beautiful appearance in the spring, when in full flower, but the fruit is not eatable, and the wood is of little value.

---

# FAMILY CONIFERÆ, DIAGRAM 16.

This family may be known at once by the very peculiar ap-pearance of the trees which compose it, such as the *pine*, the *fir*, the *cedar*, the *birch*, and the *juniper*. They resemble no other plants. Their leaves are hard, slender, with parallel fibres, like the leaves of monocotyledons; and they do not fall off annually; in other words, the trees of this family are all *evergreens*.

The name of the family is derived from the fruit, which is cone-shaped, and is commonly called *pine-cone*. The flowers are inconspicuous, as in the Cupuliferæ, the sexes are separate, the male flowers form clusters, and the female flowers are likewise often arranged in clusters. The latter increase and become pine-cones ; their scales thicken, and inside each we find a fruit furnished with a membrane like a wing. The male clusters are often crowded together, and yield an abundant pollen which forms a yellow dust. If we gather a branch of fir in spring, which is thus loaded with clusters, we shake off much of this dust, which the wind carries to a great distance. If all the pollen of a pine forest is carried away by a gust of wind, it forms a real cloud when it descends upon the country at a distance. As this pollen is yellow, and burns easily, it has sometimes been mistaken for sulphur.

Plants of the family Coniferæ nearly always grow on mountains, or in dry countries, even on the sands of the sea-shore. Their timber is specially adapted for ship building, and they also yield resin, tar, and pitch. Fir trees are sometimes planted on sand-dunes near the sea, that their roots may hold the sand together and prevent its spreading further inland, and making the neighbouring country sterile.

The *pinaster* is a cultivated plant in England. On the continent it grows either on mountains, or by the sea-side; its leaves are long, and inserted in pairs into a common sheath. The scales of the cone are thick. The pinaster is cultivated abroad for the sake of the resin which it yields while living, and the planks which it makes when felled. It is not used until it is of about twenty or thirty years growth, and then the collectors of resin make an incision through the bark about a foot long, and the resin flows from under the bark at the top of the wound. Every week they stimulate it by removing a small layer of wood, and the resin which was a little slackened begins to flow better again. It is received in earthern jars which are put under the incision at the foot of the tree. It is gathered from the month of May to the end of September. When the turpentine, which is liquid, has been extracted from it, the resin, which is hard and brittle, remains.

The *firs* may be known by having their leaves arranged like the teeth of a comb ; the cones are cylindrical, and formed of slender scales.   The firs are large trees which only grow on high mountains.   Their trunk is always very straight, and is used to make masts for ships.   The fir also yields resin, but not abundantly, and they are contented to collect what flows naturally and which is found in large lumps on the trunk.   This resin is more esteemed than that of the pine.   But the best is that which is called *Venice turpentine*, and is obtained from the larch tree.

The *cedar* is a beautiful tree which was originally brought from Mount Lebanon, but which is now frequently grown for ornament, as it is very hardy.   Its wood is much used by joiners, and has a pleasant smell.   It is also generally employed for making lead-pencils.

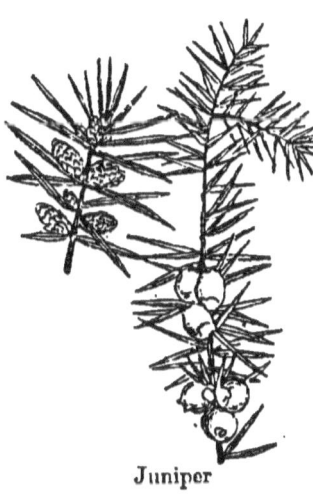

The *yew* and the *juniper* do not bear cones like other coniferæ, but berries, and the berries of the yew are red.   Yew-wood is also highly valued for its elasticity, and it is used for making bows.

Juniper

# CLASS OF MONOCOTYLEDONS

## OR PLANTS THE SEED OF WHICH HAS ONLY ONE COTYLEDON.

---

Some of the families of monocotyledons have very beautifully coloured flowers, as the lily, the flag ¸and the tulip; and other families, such as the grasses, have inconspicuous flowers. These latter are generally the most useful families to man, as among the dicotyledons.

---

## FAMILY LILIACEÆ.

Lily.

It includes a great number of ornamental plants like the *tulip*, the *lily*, and the *hyacinth*, and other plants which are used for food or condiments, such as the *shalot*, *leek*, *garlic*, and *onion*. The *aloe* also belongs to the liliaceæ. The fibres of their leaves are all parallel, a character common to all the monocotyledons. The root is a bulb from which a stalk grows up annually and dies in autumn. The liliaceæ have no calyx. They have a beautiful corolla with six divisions; there are six stamens, and the fruit is a capsule with three valves and three divisions. The corolla is inserted below the ovary.

The *tulip* is remarkable for its gay colours, but it is of no particular use, nor is the lily.

The *aloe* is an African plant, and largely cultivated in the West Indies, which yields a resin which is used in medicine as a purgative, and its leaves are very large, and yield a textile substance composed of very coarse, but very strong threads : ropes, and cordage are made of it.

The *leek, garlic, onion* and *shalot* have flowers arranged in an umbel, as in the Umbelliferæ, but the stalk is only divided once. The flower, as in the lily, has six divisions and six stamens. In all these plants it is either the bulb or else the base of the leaves which is used for flavouring.

The name of the family Liliaceæ is derived from the Latin name of the lily.

---

# FAMILY IRIDACEÆ.

The Iridaceæ include odoriferous flowers, like the *flag* and the *saffron*. Like the Liliaceæ, the flowers of this small family have a beautiful corolla, and no calyx. They have six divisions ; three outer and three inner. There are only three stamens. In the flag, they are hidden by the pistil, and terminated by three very large divisions which at first appear like petals. One important difference between the iridaceæ and the liliaceæ is that the flower of the former is inserted on the ovary, instead of being inserted below it.

Saffron.

The pistils of the saffron flower are gathered to make a red colour. They are also used in confectionery and are thought to improve the health of caged birds when put into their drinking water. They are dried and powdered before being used. They have a very penetrating odour which can cause illness, and that of the flag is also sometimes attended with very serious effects. The root of the flag is odoriferous, and is used after being powdered.

# FAMILY AMARYLLIDACEÆ.

This is another small family which also comprises handsome plants, like the *daffodil* and *narcissus*. Their corollas have six  divisions like those of the Iridaceæ which are also inserted on the ovary, but there are six stamens. The daffodil and narcissus, which are common in meadows in many parts of England, and are

Narcissus.                Agave.

still more frequently grown in our gardens on account of their beautiful white and yellow flowers, are very poisonous plants, and their odour, like that of most strong-scented flowers, will make some persons ill. The *agave* or *American aloe* belongs to this family, and bears leaves very similar to those of the aloe, which yield a thread of great fineness and strength.

Palm tree.

# FAMILY PALMÆ.

Palm trees are all inhabitants of warm countries, and no species is found further north then the extreme south of Europe. They sometimes grow to a very great height. We have mentioned already (see p. 169) that their trunk does not increase in thickness with age, but only in height. It is surmounted by a crown of beautiful leaves which make it bend before the slightest wind. The palms yield very various kinds of fruit, dates and cocoa-nuts for example.

# FAMILY GRAMINEÆ.

This family is indisputably the most numerous of the vegetable kingdom. It includes all the plants which are commonly called *grasses*, and many others also. It furnishes man with a large proportion of his food in the form of *cereals*, which are all grasses. The *sugar-cane* is also a plant of this family.

Most of the Gramineæ are annuals which must be planted annually, as they die every year. Their stalk is not like that of other plants, but is hollow, with knots at different intervals. A leaf grows from each knot, which at first surrounds the stalk, and then spreads out from it. The flower of the grasses is inconspicuous. Its centre is composed of a pistil with a bifurcated extremity resembling two small and very light feathers. Round the pistil hang three stamens, the pollen-sacs of which are supported at the end of very fine threads. The pistils and stamens compose the whole flower, they issue from more or less numerous scales which are called chaff, and which remain round the corn.

There is only one for each flower, and corn is threshed in order
to free it from the chaff which envelopes it.

Wheat is sown either at the beginning of winter or at the beginning of spring.  When sown in autumn, it sprouts and resists
the cold, but does not grow through the winter, and only begins
to grow again on the return of warm weather.   Each grain of
wheat produces sometimes more or sometimes fewer stalks, and
consequently ears.   After the plant has flowered, and the corn
has acquired its full growth, the stalk and ear begin to turn
yellow, or in other words, to die.   The wheat is then reaped,
and afterwards threshed to separate the corn from the chaff.

There are many different varieties of wheat suitable for different soils and situations.   Some are much more hardy than
others.   The less hardy kinds yield a floury grain which can be
easily reduced to powder, and are the best for bread making.
The hardy kinds are smaller, more horny, and are more easily
broken than ground; they yield a flour which makes very nutritious bread, but not of a fine colour; it is well adapted for pastry.

In order to make bread, the wheat is put into a mill, where it
passes between two millstones, one of which revolves, which
grind it, but the flour in this condition is not yet sufficiently
prepared; it is mixed with the bran which proceeds from the
outer skin of the corn.   This is removed by passing the flour
through a very fine sieve.   Wheat, thus converted into flour,
will keep for a long time in a dry place.   When it is required
for use, it is kneaded with water and salt, but if cooked thus, it
yields a heavy compact dough which is not fit to eat. It is necessary for the bread to *rise* when it bakes, and we must therefore add
to it a little *yeast*, a microscopic fungus which becomes developed
in the course of the fermentation of beer, and which is found in
the form of a whitish scum in the vats where beer is made.
When this is done the dough will rise in baking.

Bread is one of the best foods known, and is the more nourishing in an inverse proportion to its whiteness, but it is by no
means indispensable to life; and potatoes, beans, rice, or meat
can be substituted.   Bread is the principal article of food in
some countries, while the inhabitants of others eat little or none,

and supply its place with other food. *Maize*, which is much grown in America and in the south of Europe, is one of the Gramineæ, but differs somewhat from the cereals; the flowers have only one sex, and form two ears on each plant. The ear

of male flowers is at the very top of the stalk. That of female flowers is lower, and nothing can at first be seen but a tuft of large pistils, which unfold from the leaves. The ripe grains form a compact ear which is stripped to make flour. It is chiefly used in England for making puddings, &c., but in America the grain is cooked in a great variety of ways.

Ear of maize.

The sugar-cane is another plant of this family, which is largely cultivated in the West Indies. All the sugar used in Europe was obtained from it until the beet-root begun to be cultivated for the production of sugar. The stalk of the sugar-cane is nearly as large as the arm of a

young child, the knots are very close together, and the whole interior is filled with abundance of a sweet sap. When it is time to gather them, the canes are pulled up, stripped of their leaves, and passed between heavy rollers, which crush them. The juice thus obtained is evaporated in ovens, and the residue is the *raw sugar* which is imported into Europe. To make white sugar, it must be refined. *Molasses* is the refuse sap which will not crystallise after evaporation. The remains of the crushed canes are not wasted, but are fermented to make rum.

Sugar-cane.

The maize and sugar-cane are very much larger grasses than those of our fields and our cereals, but there are others which greatly surpass these in size, and which reach the size of trees. The *bamboos* which grow in warm countries are really gigantic grasses. They are sometimes as large as the arm or the leg; as the space between the knots is hollow, it is enough to cut the stalk between each knot to make household utensils.

# CLASS OF ACOTYLEDONS,

## OR PLANTS WITH NO COTYLEDON.

All the plants which still remain to be noticed, and which form the class of acotyledons, are distinguished from the others, not only because the embryo has no cotyledons, but because the plants themselves have no flowers. They never bear either pistils or stamens. At the proper season, seeds appear at some part of the plant, but do not succeed to flowers. These seeds themselves are most frequently of extraordinary minuteness, so that they resemble dust, like the pollen of coniferæ. They are so different from all other seeds that they have received a special name, and are called *spores*. Some families of the acotyledons still resemble other plants to some extent by their greenness and a kind of foliage, but there are others, such as mould and fungi, which are entirely different from ordinary plants.

# FAMILY OF FERNS.

Tree-fern.

Ferns are green, and resemble other plants. The spores are developed under the leaves in small clusters of variable form. Sometimes they are long and narrow, and sometimes round or bean-shaped. In one beautiful fern called the maiden-hair, the stalk of which is black and slender, the spores are placed under the very edges of the leaves which seem to be folded over to cover them. In the so-called *flowering fern*, the spores are arranged in a kind of stem, but its resemblance to a flowering plant is only apparent, and not real.

Our British ferns are all plants of moderate size, but in hot countries they grow to a great height, and resemble the palms, both because their stem grows in length without increasing in breadth, and because they are likewise surmounted with a crown of large leaves.

# FAMILY OF MOSSES.

The mosses, like the ferns, resemble other plants in external appearance ; they are green, and have a woody stalk, one might say, like those of very small trees. The spores of mosses are not developed under the leaves, as in ferns, but in special and very elegant organs. When we look at mosses at the season that they fructify, we notice straight slender filaments projecting

from their foliage, and terminating in a knob of very complicated construction at their extremity. It at first appears covered
with a small hat formed of fibres like a straw roof; it is pointed
and hangs over the sides of the knob. If we remove it, we find
beneath it a capsule closed by a cover. It is shaped like a wine
glass, and the cover itself has sometimes a kind of small button
in the middle. It must be raised in order to see the spores arranged in the little cup. When they are ripe, the hat and cover
fall off and allow them to escape.

# FAMILY OF FUNGI.

We no longer find the ordinary appearance of plants, either
in this family, or the two following. Fungi, or at least the
most familiar kinds, have a well known appearance, but a number of other plants, such as the moulds, must be arranged in
this family, which never present the umbrella-shape of ordinary
fungi.

This *umbrella* or *cap* of the fungus, is supported on a stalk of
variable thickness. Sometimes the cap has slender pendant
plates, called gills, below, and the fungus then belongs to the
genus *Agaricus*, which includes the true *mushroom*; and sometimes
the under-surface of the cap exhibits only a multitude of tubes
crowded together, and open at the lower extremity, and the fungus then belongs to the genus *Boletus*, or *Ceps*. But there are
many others, such as the *morels*, the *puff-balls*, and the *truffles*,
which have quite a different appearance.

The stalk of the mushroom often has a kind of collar or ring
at about two-thirds of its height. The stalk sometimes grows
directly out of the ground, and sometimes out of a kind of

bladder, which seems to burst to let it come out. This bladder
is called the volva. It is an important peculiarity to know, because it enables us, in some cases, to distinguish the species of
fungi.

Many fungi are poisonous, and others have a pleasant taste.
It is very necessary to remember that no one can distinguish the
wholesome and poisonous kinds without assistance; for they
cannot be separated either by their odour, or by the places
where they grow. Which are good and which are bad can only
be learned by long experience, or by the aid of a person who is
thoroughly well acquainted with the species. But in any case,
it is necessary not to decide at a glance, but carefully to examine
the colour of the fungus, to make sure that it is really the kind
which we imagine. If, after eating fungi, the least inconvenience is felt, an emetic should be taken at once, and a doctor
called in.

The only fungi commonly eaten in England are one or two
species of mushrooms, which can only be confounded with others
by gross carelessness : but many other fungi are commonly eaten
in France and Italy ; and it may be mentioned that the puffballs
are unmistakeable, and perfectly wholesome.

Mushrooms grow very fast, and it is common to find a quantity
in a field where there were none the evening before. They are
grown on hotbeds in a cellar, or in a damp, dark place. When
the hotbed on which they grow is removed, whitish filaments
are seen which are called *mushroom spawn* or *mycelium*. Mushrooms
only grow from this, and it is put into the beds to sow them. Most
fungi are eaten by the larvæ of insects, which are very fond of
them. When the puffballs are ripe they become dry and brown,
and when pressed, burst, and clouds of dust entirely formed of
spores, fly up. Dried puffball is a useful remedy for stopping
bleeding.

*Mould* is composed of different kinds of very small fungi,
wholly formed of slender filaments analogous to those which
form mushroom spawn. When they are ripe and filled with
spores they assume a greenish tint.

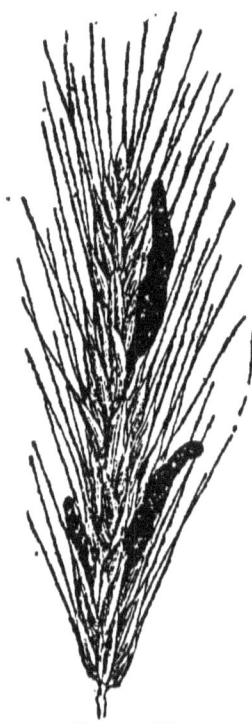

Ergot of Rye

There are many kinds of fungi similar to mould, but still smaller, which are invisible to the naked eye, and can only be detected by their ravages. The *oidium* which attacks the vine is a fungus of this description; and the disease called *muscardine* which attacks silkworms, is caused by another; the potatoe disease is either caused, or frequently accompanied by, a similar fungus; another produces a disease called *fumagine* in the olive; and yet others produce *mildew*, *smut*, and *ergot*, in cereals. The ergot chiefly attacks rye, when instead of the ordinary grain, we see large black horny grains developed, shaped something like a cock's spur. Ergot is poisonous, but is of great use in medicine. Man himself is subject to diseases caused by the presence of fungi of this kind, such as the *thrush* and *ringworm*. In the thrush, the white points which form on the tongue and inside the mouth of children, are caused by the presence of a fungus. In the ringworm it is also a fungus which form the yellow crusts shaped like buttons, hollow in the middle, which grow at the roots of the hair.

We may add that the yeast which we have mentioned as found in beer vats is also a fungus of this kind.

---

# FAMILY OF LICHENS.

The *lichens* resemble fungi, but instead of being moist, like the latter, are dry. They often resemble parchment, and we should scarcely take them for living plants if we did not see them grow with time.   They are generally of a grey colour, and sometimes grow on the trunks of trees, and sometimes on walls and slates. They require very little moisture.   Some lichens appear like large pendant filaments, which sometimes  cover over fir-trees, . and finally destroy them.

---

# FAMILY OF ALGÆ.

The family of algæ includes, firstly, the green filaments which are  found  in stagnant waters, and secondly, the marine plants, which are sometimes of a beautiful green, and sometimes of a brown or red; they cover the rocks, and it is a pleasant amuse-ment to collect and dry them.   But they must be allowed to lie for some time in fresh water, till all the salt which they contain has been washed out, or else they will not dry.

The marine algæ, or *seaweeds*, are useful for many purposes; firstly, they form an excellent manure when collected and spread upon the land; and secondly, soda, iodine, and other chemicals which are very useful in manufactures and medicine are pro-cured from them.   Soda is obtained by drying seaweeds in large heaps, and then burning them and washing the ashes. The water dissolves the soda, and on evaporation, the soda is deposited in crystals.

Some species of seaweeds are used for food, such as the *dulse,* which is eaten  boiled, as a kind of vegetable, and the *Carrageen Moss* which is used to make a kind of jelly.

The longest of all known plants are seaweeds; and one species which abounds in the Antarctic seas is said to grow to the enor-mous length of 6C0 feet.

# MINERAL KINGDOM

Tho Mineral Kingdom includes, as wo havo said, all substances which aro not organic. Tho different kinds of stones
and metals belong to the mineral kingdom, as well as water;
and the gases which are mingled in the atmosphere.

If we descend into a quarry or cutting, and look at tho sides,
the soil generally appears to be formed of different kinds of
earths or rocks arranged one above another. These aro called
*strata.* Sometimes theso strata are horizontal, and at other
times they are more or less slanting. There aro some rocks,
however, which do not prosent this stratified appearance, and
which simply form large masses, like a single block. This is
more ospecially the case in countries where granite is found;
but still two layers can be distinguished, for the granite is
nearly always covered with a layer of vegetable earth. The
formations when the strata are arranged one above another are
called *sedimentary,* and the others *primitive* formations.

Primitive formations are so callod beeause they are thoso
which were most anciently formed. The sedimentary formations
have been deposited above them by tho agency of water.

In sedimentary formations, the strata are often very dissimilar
to each other. They also frequently contain the imprints of
animals or plants, or else bones and petrified tooth. These remains of beings which lived formerly aro called *fossils.* In
some places there is an innumerable quantity of them, but in

others they are rare. When any are discovered which do not
resemble those which are commonly observed, or when they are
found in formations where they are very rare, they ought always
to be preserved, when possible, to be shown to persons who
know and can appreciate their scientific value.

*Mountains.*—The central mass of mountains is generally composed of granite, while sedimentary strata are found on their
flanks and in their valleys. The snow and rain which accumulate on the highest mountains sometimes form great masses of
ice, which may be several miles long and broad, and are called
*glaciers.* The thickness of these masses of ice is sometimes considerable, and they contain great cracks called *crevasses* into
which there is danger of falling when they are hidden by newly
fallen snow.

While the fall of snow and rain increase the glacier, it par-
tially melts in the sun, and rivers which run into the valleys
always flow from the foot of a glacier. In spring when the
rain which has fallen on the mountains melts away, these rivers
are transformed into torrents. Their force then is irresistable,
and they carry away with them earth, pebbles, and rocks from
the highest parts of the valleys, and carry all this mass into
lower lying districts, where they form in time, immense fertile
plains. Consequently elevated countries are always being more
and more worn away by the action of the rain, and the melting
of the snow; and this action is called *erosion.* The waters of
heaven thus remove, little by little, and by piecemeal, the soil
from the sides of mountains, just as the waters of the sea
gradually eat away the land on many coasts by another kind of
erosion.

*Springs.*—The water which results from the melting of snow
and ice, is not the only water which flows on the surface of the
earth. The greater part is supplied from springs which appear
to rise from beneath beds of earth. The fact is that all the rain
which falls does not flow into the brooks, nor dry up in the air.
When much rain falls, it always penetrates and infiltrates into
the ground, till it reaches a layer which it is unable to traverse.

either because the rock is too hard, or because it is clay, and all this water flows at the surface of the impenetrable layer which stops it, and flows out in the neighbourhood forming a spring.

*Warm Springs.* — Warm springs, possessing medical properties for the cure of diseases, are often met with in mountainous countries. They are called *mineral waters*, or *thermal springs*, thermal being derived from a Greek word meaning warm. Sometimes these waters contain a large quantity of sulphur, and smell strongly of rotten eggs.

*Wells, Artesian Wells.*—The wells which are dug to obtain water are designed to strike one of these springs of subterranean water which is supposed to exist at a moderate depth. When it is found, the water is soon to filter into the well from all sides

Subterra-
nean
Spring

Section of strata pierced by Artesian wells.

of its surface, and it is raised from the bottom by buckets or a pump.

But it may also happen that a spring of water enters between two layers of impenetrable soil, resting on the slope of a mountain, and then one part of the spring of water will soon be much higher than the other, and will form a kind of raised reservoir, like those constructed to give more force to water, and make it rise in a fountain in ornamental basins. Artesian wells are designed to pierce to a great depth in the ground, to attempt to strike a spring of this kind, for then there is no occasion to draw the water, for it will rise to the surface of the ground of itself. The name of these wells comes from the district of Artois, where the first were dug.

*Lakes.*—When several springs unite, they form streams, which in their turn unite to form a river, which flows to the sea where the formation of the country permits it. But if this is not the case, and the formation of the country opposes its course, the waters accumulate, and form a *lake* or a *pool*. When the water from rain or from a river is discharged into broad plains where it can neither flow into the sea, nor accumulate to form a lake, these plains become marshes and peat bogs.

*Banks and Cliffs.*—If rain water daily tends to remove pebbles

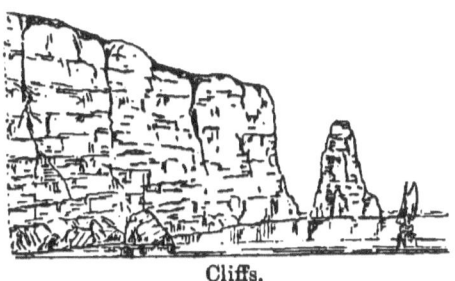

from the mountains, and to deposit gravel in low lying plains, the sea also plays a double part on its banks. It encroaches on the land in some places, and is encroached upon by the

Cliffs.

land in others. This is especially noticeable at the mouths of rivers. The earth and gravel carried down by the current after rain gradually accumulates at the mouths of rivers and streams and forms flats which advance further and further into the sea. The sea in its turn throws up heaps of sand on some coasts, while at other points it wears away its banks, and advances into the land. When this is elevated a cliff is the result.

*Volcanoes.*—There are mountains in some countries with a large hole in the summit called a *crater*, whence issue stones,

Volcano in eruption.

dust, smoke, and flames. These mountains are called *volcanoes.*

Sometimes the side of the volcano opens, and discharges incandescent *lava*, like the molten metal that flows from a furnace. This lava forms a true river of fire, which flows upon the surface of the ground, and consumes everything in its track, but it generally advances very slowly. There are no longer any volcanoes in England or France, but there were formerly many. The best known of these extinct volcanoes are those of Auvergne, in France, which no longer discharge smoke, or flames or lava,

Extinct volcanoes of Auvergne.

but which were formerly active, and discharged lava-torrents which can still be traced.

*Earthquakes* are tolerably frequent in the neighbourhood of volcanoes, though they may also occur in countries which are not volcanic. These are shakings of the ground which, when sufficiently violent, cause great catastrophes by overturning houses. But earthquakes are fortunately very uncommon, and by no means severe in England, for the very reason that there are no volcanoes nearer than Iceland and Italy.

*Amospheric air.*—The air which surrounds us, and which we breathe, is also a mineral substance in a gaseous state. We have mentioned its composition and properties at pages 11 and 12.

# INDUSTRIAL MINERALS, DIAGRAM 19.

After having briefly described the most remarkable phenomena of the earth, we will notice the principal useful substances which man has found in it. Those which he obtains from the mineral kingdom are not less numerous or important than those derived from the animal and vegetable kingdoms.

*Granite.*—The primitive formations are chiefly composed of granite. It is found at the surface of the ground in many places, only covered by a thin layer of vegetable soil.

Granite is generally considered the hardest and most durable of stones. There are certainly some kinds of granite which possess these qualities, but all do not possess them in the same degree, and there are some kinds of granite which deteriorate very easily. Hard granite is used for engineering purposes, but does not answer so well for architecture. It is sometimes rose-coloured, and sometimes bluish or black. The working of this stone is always somewhat difficult, firstly on account of its hardness, and secondly because it exists like all primitive rocks, in enormous masses which are more difficult to deal with than rocks, which are perhaps quite as hard, but which are arranged in layers one above another.

Fossils are never found in granite.

*Pumice-stone.*—Pumice-stone is a volcanic product, and is only found where volcanoes exist or have existed. It is a stone remarkable for its porous structure, which causes it to float in water on account of the air contained in its cavities. Pumice-stone is very brittle and friable, but at the same time very hard. It is reduced to powder, and used in industry to polish wood and ivory, as well as leather and parchment.

*Sulphur* is another volcanic product which is found under the same circumstances as pumice-stone, but sulphur is of much more extensive use in industry. It is employed to make gunpowder, mixed with charcoal and saltpetre. It is used for matches, and is also employed in the manufacture of *sulphuric acid* or *oil of vitriol*, great quantities of which are used in industry. Lastly, powdered sulphur is also used by agriculturists to destroy microscopic fungi on plants. The sulphur gathered near volcanoes is sometimes very pure, and is then called *native sulphur*. But it is generally mixed with earth, and must be purified before it is fit for use. Sulphur burns in air with a blue flame, but when it is heated out of contact with air, it volatilises and becomes deposited in the form of a powder which is called *flowers of sulphur*.  This is what is used in agriculture.

*Slates and schists.*—All laminated rocks are called in a general way schists.  To extract slate, large square blocks of suitable size are first detached in the quarry ; it is then carried to the workshops, where workmen called splitters divide it into sheets of variable thickness, and these are the slates which are used for roofing houses.  Slates are also used to make school writing tablets, billiard tables, and whetstones.

*Coal.*—Coal is one of the most precious productions of the earth.  It is generally found at great depths, whence it must be procured by mining, and the aid of powerful machinery.

The districts where coal is found are called *coal fields*.  Coal belongs, like the schists, to the lowest sedimentary deposits.  It is generally arranged in thin layers, near each other, and very

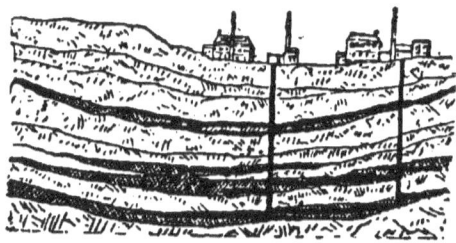

Coal-mine.

slanting.  When a shaft has been dug deep enough to reach these layers, galleries are pierced into them by means of which the coal is brought to the foot of the shaft, from whence it is raised by machinery.  These galleries are generally very narrow, and just large enough to allow a small wagon to pass.  The coal and rock are then separated.

Coal-mines are subject to special dangers which do not exist in other mines to the same extent.  Water is often abundant in them, and must be pumped out day and night to allow the work to proceed, for if the pumps should stop for only a minute, it might happen that the galleries would be inundated, and work become impossible.  Another danger is *fire-damp*.  The coal in some mines emits a gas which explodes as soon as it touches a light, shatters the works, and kills the miners. This danger is avoided by using lamps, the flame of which is protected by wire gauze, and the fire-damp burns within the gauze, but does not communicate the flame to that without. Another danger is *choke-damp*, a gas likewise emitted by the coal, which does not explode like fire-damp, but which suffocates the miners.

Another danger is that of a coal mine taking fire, which is less dangerous to the life of the miners, but which leads to enormous losses. Whenever it occurs, the mine must be closed immediately, and several years must sometimes elapse before work can be resumed.

Coal is a source of immense wealth to the countries where it occurs. It is not only used in fire-places, for its principal use is to heat the water in the boilers of steam-engines. But coal is also used for many other purposes. The gas which is used to light our streets and houses is made of it. For this purpose it is heated in iron retorts out of contact with air, which is called the destructive distillation of coal. The gas which it produces is collected in large receptacles called *gasometers*, and *coke* is found in the retorts when they are opened. Other products besides coal gas are obtained by the distillation of coal, such as *sal-ammonic* and coal tar. When coal tar is distilled anew, a number of useful substances are extracted from it, such as the beautiful dyes called *mauve, magenta, aniline*, &c.

*Bituminous schists.*—These are laminated rocks like slate, but mixed with a large quantity of bitumen. When this stone is distilled, *rock-oil* for lamps is obtained from it, as well as substances analogous to coal tar.

*Bitumen* or *Asphalte.*—In some countries this substance flows from strata formed of bituminous schists, either alone or mixed with water. At other times it is found solid, but it then melts at a low temperature. On distilling it, oils similar to rock oil, or to petroleum are obtained.

*Petroleum* is now in universal use. Its name is derived from two words meaning mineral oil. The use of petroleum has become much more general since the discovery in America of true subterranean springs of petroleum as abundant as springs of water.  When wells are dug, the petroleum immediately flows in abundance, and sometimes rushes up to a considerable height, ike the water of an artesian well.

*Graphite.*—Graphite, also called *plumbago* or *black-lead*, is found in granitic formations, in small layers or masses.  The last term is improper, as it does not contain a trace of lead, but is a kind of charcoal which burns with difficulty.  When it is extracted in blocks of sufficient size, it is sown into small square sticks, which are then enclosed in two pieces of wood to make pencils.

When the graphite is not of sufficiently good quality, it is ground and made into a paste which is allowed to dry, and from which the leads for the pencils are cut.  Powdered plumbago is used to blacken and polish grates and fire irons.

The most important and best known mine of graphite is at Borrowdale in Cumberland.

*Limestones.*—All soft or hard rocks are called by this name which yield quick-lime when calcined in the fire.  Limestones have another peculiarity.  If a drop of vinegar or any acid is poured on a rock of this nature, it immediately produces an *effervescence* of small bubbles of carbonic acid gas.

Limestones may exhibit every degree of hardness, from that of chalk to that of marble, and they may be of the most different colours.  There are black, yellow, red, white, and grey marbles, and othsrs which are veined with several colours.

Marbles, on account of their hardness and beauty, are the stones which are generally employed for monuments.  They are generally very expensive, especially white marble, which is used for carving statues and busts.

All limestones do not possess the hardness and beauty of marble, but these stones are used almost everywhere for building and stone-cutting.  They contain many fossils.

*Chalk* is a very soft and white limestone, which becomes a paste in water.  When this paste is sifted, it is dried, and called *whitening*.

There is a very finely-grained grey or yellowish limestone which is used for taking impressions. A drawing is made on these stones with a particular kind of ink, and by pressing the design against a sheet of white paper, we can take as many impressions as we wish. This process is called *lithography*, from two Greek words which mean *writing on stone*; and the figures in this book have been produced by lithography.

We have mentioned that all limestones yield quick-lime when heated to redness, and this operation is performed in *lime-kilns*. When quick-lime is thrown into water, it grows hot, and forms a paste with it which is mixed with sand to make mortar. There are three kinds of lime; *fat lime, poor lime*, and *hydraulic lime*.

Fat lime is produced by the hardest limestones. A large proportion of sand can be mixed with it to make mortar, and it is therefore economical, but the mortar is not very firm.

Poor lime is generally not so white, and much more of it must be employed to make mortar, but it holds much better.

Lastly, hydraulic lime, or Roman cement, is made of limestones which contain a large proportion of clay. It is then enough to mix it with water to produce a paste which immediately becomes very hard, so it is used to construct works which are to be submerged. When hydraulic lime is placed in water it becomes harder and harder.

*Sands and sandstones.*—There are entire layers of sand in the earth similar to that on the sea-shore. It is sometimes white, sometimes bluish, and at other times red, in consequence of containing iron.

*Tripoli* is an exceedingly fine and hard sand which is used to polish metals.

*Sandstone* is entirely formed of conglomerated grains of sand. Sandstone is sometimes friable, and easily disintegrated, and at other times it forms a very durable stone which is used for paving the streets of towns.

Silex or flint is formed of the same substance as grains of sand. It is generally blackish or grey, but sometimes reddish. It breaks like glass, and is sometimes semitransparent. When struck with iron, sparks appear, which are particles of iron chipped off and ignited by the concussion.

*Mill-stone grit* is another kind of silex which is generally of a white or reddish colour, and is hollowed with a number of large and small cavities. In spite of the presence of all these holes, millstone grit is an exceedingly durable stone, with which foundations and buildings which are · required to possess unusual solidity are constructed. Millstone grit is so called because it is also used to make millstones. For this purpose stones must always be chosen which are both hard and full of holes, so as to crush and bruise the corn better. Millstone grit answers these conditions very well, but it is very rare to find quarries in which millstones can be cut in one block, so they are made of several pieces fitted together, and joined by cement. The millstone is then strongly bound with iron, and allowed to dry for a very long time before being used. These millstones are as solid as if they were made out of a single piece.

*Rock-crystal, agate, glass.*—Rock-crystal and agate are also formed of the same substance as sand, sandstone, and millstone. This substance is silex, flint, or quartz. When it is perfectly pure, it is also perfectly transparent, and forms rock-crystal. In agates, the quartz is slightly coloured, or traversed by veins of different colours. Rock-crystal and agates are extremely hard, and are frequently used to make ornamental articles.

Glass is made of sand and soda, which is melted together at an intense heat. Fine glass is often called *crystal.* It is more sonorous, and is cut easier than glass. It is obtained by adding a definite quantity of litharge to melted glass.

*Clays.*—Clays are earths of exceedingly fine grain, which form, when mixed with water, an adhesive paste, which can be worked in different ways. They are composed of more or less impure alumina. Clays are sometimes bluish, yellowish, or red. Though they can be mixed with water, water cannot easily penetrate them, and we therefore see the rain remain on the surface of the ground for a long time in clayey districts.

Bricks, and the different kinds of ¡crockery and earthenware,

are made of the paste formed by the mixture of clay and water . Porcelain is simply made of a white kind of clay called *kaolin.* All bricks and earthenware must be baked before they are sufficiently solid to be fit for use.

*Fuller's earth* is a greenish clay, greasy to the touch, which dissolves in water, and makes it soapy. It is used to remove the oil from cloth, for it is necessary to apply oil to wool before it can be spun and woven.

*Gypsum, or plaster of Paris.*—Gypsum is a stone which has much external resemblance to limestone, but it does not effervesce with acids, and does not yield quicklime when burnt, but *plaster.* Sometimes gypsum is found in the form of large crystals shaped like spear-heads, which also yield plaster when heated. This does not require so intense a heat as the manufacture of lime. When gypsum is taken from the kiln, it is easily reduced to powder which again becomes solid when mixed with water. If plaster has been exposed to the air before using it, it is no longer serviceable, because it has absorbed moisture from the atmosphere. Plaster is used either to join brickwork or masonry, or to make ornaments and stucco on walls. It is also run into moulds, and is taken out with the desired form. This process is called *moulding,* and is used to reproduce statues and busts.

*Rock salt.*—A large part of the salt which we use is collected by evaporating sea water in shallow pools. This is called *sea-salt,* or *bay-salt.* But salt is also found in the earth in layers. It is sometimes deposited in thick beds like rock, from which it is hewn like stone. It is then very white and transparent, and called *rock salt.* At other times the salt is mixed with clay or sand in the earth, and in this case water is allowed to flow into the mines, which dissolves the salt, and afterwards deposits it on evaporation.

Salt is not only very useful for food, for salting provisions, making bread, &c., but is also used to make *soda,* a substance very useful in industry. We have mentioned that it can be extracted from sea-weeds, but the greater part is manufactured directly from salt.

*Diamonds and precious stones.*—Most precious stones are only valuable and esteemed on account of their rarity. But the diamond is the hardest known substance, and scratches all others. It is therefore used to cut glass. There are no diamonds found in Europe, but they are generally found in gold producing countries, such as the East Indies, America, and South Africa. They are generally found in the gravel of river-beds, but they must be cut in facets to give them lustre.

*Peat.*—Peat is not, strictly speaking, a mineral. It is a deposit of dead plants which collect in the waters of marshes, and on clayey flats which have no outlet for the rain. The plants and mosses grow one above another, and finally form a compost which burns readily when dried. This compost is then removed in clods and put to dry in the sun. They shrink very much, and then form a good combustible, which has only the inconvenience of producing much ash.

*Guano.*—Guano is only found in the Chinchas Islands, off Peru, and is brought from thence to be used as manure. It is a yellowish earth with a very strong odour. Remains of the feathers and bones of birds are often found in it, and it is believed to be chiefly composed of the dung of sea-fowl, which have frequented these islands for a long period. Guano is one of the best manures known, but it must be mixed with earth or other substances before it is used, or else it would be too strong, and destroy the crops instead of improving them.

# ORES.

Those productions of the earth from which metals are extracted are called ores. It is rare to find metals existing pure in a natural state, although some are met with in this condition. Generally they are completely unrecognisable, and are only obtained by more or less complicated processes.      *Metallurgy* is the name given to these operations.

Ores sometimes form considerable deposits, but they are generally arranged in thin layers called *veins or lodes*. These veins are often several leagues in length, by a breadth which does not exceed a few inches. Most veins form hard masses, which are worked by being *blasted* with powder. The mines which are dug to procure them are not generally subject to the same dangers as coal mines, but are often difficult to work on account of the hardness of the rock in which the veins are embedded.

*Iron ore.*—This is often met with at the surface of the ground, and is nearly always of a red colour similar to that of rust. To obtain the iron, the ore is thrown into very hot furnaces, called *smelting furnaces*. Coal is thrown in at the same time, and the molten metal is collected at the bottom of the furnace, where it flows into trenches of sand, and cools in masses which are called *pig-iron* or *cast-iron*. This cast-iron must be melted again before it can be used. It is then made into a great many articles such as grates and kettles, and it is also made into stoves, but these ought to be used as little as possible because they are unhealthy, and may even cause serious accidents to persons who work in rooms and workshops heated by these stoves.

To work cast-iron it is put into the fire and when it is sufficiently softened it is put under an enormous hammer, the blows of which remove those substances called *scoriæ* which make it brittle and easily fusible. Iron thus purified melts with difficulty. It is not brittle, and can be forged at will, and it is called *soft iron*   When the iron ore contains a large proportion of sulphur, the iron remains brittle after having been purified.

To make steel, the iron is mixed with a proper proportion of coal, and is heated to redness. Steel is brittle, but elastic and very hard. These qualities can be increased by heating it more or less, after which it is plunged either into oil or into water, to be suddenly cooled. This process is called *tempering*. The uses of cast-iron, iron, and steel, are innumerable.

*Loadstone* is a kind of iron ore which possesses the property of attracting iron to it. When bars of soft iron are rubbed with it, they acquire the same property, and are then called *magnets*. Loadstone is not, however, the only body which thus attracts others. It is enough to rub a stick of sealing wax on cloth, and put it near the down of a feather or a very small piece of paper, to see these very light bodies attracted as is iron by the magnet. This, however, is produced by electricity, and not by magnetism.

When a piece of soft iron has thus been rendered magnetic, we find that one end attracts and the other repels iron. If the magnet is then suspended so that it can move freely, we shall find that it will move round until the attracting or *positive pole* points to the north, and the repelling or *negative pole* to the south. The *mariner's compass*, by which ships are guided across the sea, is formed of a needle which has thus been rendered magnetic, and always points north and south.

*Copper ore.*—Copper, like iron, is rarely met with in a pure state, but is nearly always mixed with sulphur. The treatment of copper ore is rather tedious. When it has been washed and crushed, it is roasted several times to burn the sulphur. It is then melted, and the copper separates. Copper is of a red colour, and is used to make boilers, saucepans, and a variety of utensils, but they ought always to be kept very clean, as otherwise *verdigris* forms in them, which is a violent poison. To avoid this inconvenience, the inside of copper vessels which are not required to be exposed to a great heat, are *tinned*, but they are useless if too much heated, as the tinning melts.

When copper and tin are mixed together in proper proportions, we obtain *bronze*, which is more durable than copper, and much less easily tarnished in the air. Bronze is used to make statues, cannons, and a great many objects which are very little injured by time.

When we add a certain quantity of zinc to copper, instead of tin, we obtain brass. These mixtures of two metals are generally called *alloys*.

*Zinc ore.*—This is also called *blende*. It is pulverised and calcined with coal to obtain zinc. As this metal is volatile, that is, is liable to be reduced to a state of vapour by the action of the fire, advantage is taken of this quality to purify it. The rough zinc is distilled out of contact with air, and distils over and condenses in drops. It must then be remelted. Zinc, like iron, can be hammered into thin sheets, which are much used in industry to cover houses, and to make pails and other utensils.

A colour called *zinc white* is also made with this metal, which unlike white lead is not dangerous to those who employ it.

*Tin ore.*—This ore is much less widely distributed than that of many metals, but it is very abundant in Cornwall. It is a very valuable metal. When pure, it is as white as silver, but quickly tarnishes. It is quite harmless, and communicates no injurious property to water. For this reason it is used to line copper vessels. It is also used to make *galvanised iron*, which is only a thin layer of sheet-iron tinned on both sides. If thus prepared, the iron does not rust as long as the layer of tin covering its surface is not worn off or removed. Tin is also used to make the solder used by plumbers, and the tin-paper in which chocolate and other substances that damp would injure are sometimes wrapped.

*Lead ore.*—To obtain lead, the ore must first be roasted several times, and the residue must then be mixed with coal and old iron, and heated, when the lead melts, and runs out. It is brilliant, like tin, when cut, and has a bluish reflection, but immediately tarnishes in the air. However, it is nearly unalterable. It makes an excellent covering for houses, and is also extremely useful for making gas-pipes and water-pipes.

However, lead is sometimes a dangerous metal, and may communicate a poisonous property to beverages. Beverages, such as wine or cider, ought not to be placed in leaden vessels, or they will be liable to produce colic in those who drink them.

*White Lead, red lead,* or minium, and *litharge,* are all compounds of lead, and the painters who use these colours are consequently liable to be attacked with colic. In many cases oxide of zinc or zinc white, which is not dangerous, can be substituted for white lead. Those who work with compounds of lead should be very careful to wash their hands before eating, and to change their working clothes frequently, and in this way they will avoid most of the dangers which are caused by the use of these substances.

*Ore of antimony.*—The metal extracted from it is white and very brilliant, with a bluish reflection. When rubbed, it emits an odour somewhat resembling that of garlic. Antimony is much used in medicine, and *tartar emetic* is composed of it. Another equally important use which it serves is to make printer's type, which is formed of an alloy of antimony and lead. Lead alone would be too soft, and antimony alone too brittle, but when mixed, a suitable alloy is obtained, which is called *type-metal.*

*Ores of gold and silver.*—Silver is often mixed with lead in ores. Gold is found pure in many countries, and is also found in some sandy districts, and in the bed of rivers. It most frequently occurs in fine particles called *gold dust.* But at other times it occurs in large masses called *nuggets.* The chief use of gold and silver is in the manufacture of money. But these metals are not very hard, and the coins would be worn out too quickly if pure metal were used. They therefore add a tenth part of copper to metal for making money, which renders it much harder and consequently more durable.

*Platinum.*—This is found in a pure state, mixed with several other metals, in the Ural Mountains, in Siberia, &c. It is white like silver, and completely unalterable, and the heat of the strongest furnace will not melt it. These qualities make it exceedingly valuable for various industrial purposes, but it is only lately that they have been able to obtain a sufficiently high temperature to melt it, which is now effected by means of the oxyhydrogen blowpipe.

*Aluminium.*—This metal is not found pure in the bosom of the earth. It is extracted by very complicated processes, from alumina, which forms the base of clay. It is white, exceedingly light, nearly unalterable, and very resistant at the highest temperatures. Much use will certainly be made of it in the arts, but it is only a few years since they have been able to prepare it.

*Mercury.*—Mercury or quicksilver is a metal which is liquid like water. It is found in a pure state in some mines, and in others it is extracted by different processes. Mercury can be boiled and reduced to a state of vapour, like water, but these vapours are dangerous, and seriously affect the health of the workpeople employed in the numerous industries where mercury is used. Mercury has the power of dissolving gold and other metals, just as water dissolves sugar. When there is a large quantity of the metal in proportion to the quantity of mercury with which it is mixed, they form a paste which is called an *amalgam*. It is an amalgam of mercury and tin which is used to form the silvering of mirrors.

***

# VEGETABLE SOIL.

This is so called because it is the soil most favourable to the growth of plants useful to man. 'It is formed naturally of sand and clay, or the detritus of rocks mixed with an uncertain proportion of organic matter yielded by all the animal and vegetable substances which rot on its surface. The presence of these organic matters is requisite to form a good vegetable soil. Dead leaves, the fragments of plants which fall on the ground, and soil drifted by the wind, all contribute to increase continually the thickness of this vegetable layer.

Many rivers overflow their banks at certain seasons of the year, and cover broad plains with their waters, which there deposit the earth which they carry with them. Soil thus formed of earth or sand brought down by the agency of water, is called an *alluvial deposit*.

Vegetable earths may be classed in four principal groups:

1st Sandy soils.

2nd. Clayey soils.

3rd. Calcareous soils.

4th. Peaty soils.

1st. Sandy soils are chiefly composed of gravel or of sand, which is sometimes very fine, and does not retain water ; these soils are liable to drought. Such soils are generally found on the shores of the sea, or of rivers. When mixed with a large proportion of decaying vegetable matter, the sandy soils form heath-soil, which may be rendered very productive by abundant watering or manuring.

2nd. The clayey earths are those where clay predominates, and agriculturists generally call them *stiff*, or *heavy soils*. They are often of a reddish colour, and may even be recognised at a distance by their appearance. When mixed with water, they form a kind of stiff paste. Clayey soils are nearly impervious to water; if sloping, they are easily cultivated, and yield large returns, but if they cover a flat country, the water cannot run off, but accumulates at the surface, and the vegetation is to some extent drowned. Then, when warm weather returns, the soil hardens, cracks, and thus tears the roots of the plants.

On the other hand, when clayey soils contain a proper proportion of sand or lime, they are excellent for agricultural purposes.

3rd. *Calcareous soils* are those which are formed by the decomposition of calcareous strata. When there is too much lime, the quality of the land is bad like clayey soils which contain too much clay, or sandy soils which contain too much sand. When there is too much lime in the soil, the surface, when moistened, cakes, and forms a crust which prevents the air from penetrating into the ground. Frost loosens the chalky soil, pulverises it, and in this condition the earth is sometimes carried away by the wind, and leaves the plants uncovered, and without proper support for the roots.

But in most cases, calcareous soils contain more or less siliceous sand or clay, and they then form excellent soils. When the chalk or limestone is simply mixed with clay, the soil is very good for the growth of corn and provender. If sand is also present, the land which is composed of the three elements, sand, clay, and chalk, also forms a soil which is favourable to the growth of most cultivated plants, and especially to the growth of trees. Calcareous soils generally yield more succulent and nourishing crops than clayey and siliceous soils; animals are generally stronger and fatter there, and their milk is richer.

4th. Peaty soils are those which contain a large proportion of more or less decomposed organic substances, and these earths, after being dried, lose a fourth of their weight when burned. They are generally of a dark colour, on account of the large proportion of organic matter which they contain.

If peaty soil does not contain a sufficient proportion of mineral substances, it is unproductive. The roots of plants cannot fix themselves in it with sufficient firmness; moreover, it dries up too rapidly, and therefore the plants which grow upon it do not always obtain sufficient moisture.

# METEORIC STONES.

This name is given to stones which sometimes fall from the sky, and they are also called aerolites. But it must not be supposed that these stones are common, and this origin has been wrongly ascribed to many fossils. True meteoric stones have always an irregular shape, and are generally of a black colour. The shooting stars which are seen at night are meteoric stones passing through the air, very few, indeed, of which, however, reach the earth in the form of stones, being nearly always reduced to an impalpable dust in the highest regions of the atmosphere.

# CONTENTS.

# ERRATA.

Page 111, line 7, for *Bombadier* read *Bombardier*.

,, 111, line 13, for *Carbai* read *Carabidæ*.

,, 111, for *Cicindelida* (under woodcut) read *Cicindela*.

,, 113, lines 5, 13, 17, for *staplylinidæ* read *staphylinidæ*.

,, 124, line 6, for *Hypenomeuta* read *Hyponomeuta*.

, 126, line 1, for *Hempitera* read *Hemiptera*.

,, 146, line 7, before *We have given* read *In the diagram*.

,, 172, line 30, for *naturity* read *maturity*.

,, 194, line 4, for *the* (at end of line) read *then*. .

,, 198, line 18, for *mose* read *most*.

,, 223, line 26, for *othsrs* read *others*.

,, 215, head-line, for *Vegetable* read *Mineral*.

The Geological specimens are on  Diagram 19.

The Mineral specimens are on Diagram 20.

www.ingramcontent.com/pod-product-compliance
Lightning Source LLC
Chambersburg PA
CBHW031422020726
47499CB00005B/1551